Geographic Data and the Law

Defining New Challenges

Geographic Data and the Law

Defining New Challenges

Edited by
Katleen Janssen and Joep Crompvoets

Leuven University Press

ISBN 978 90 5867 924 6
D/2012/1869/45
NUR: 905

Cover Design: Jurgen Leemans

GPRC
Guaranteed
Peer Reviewed
Content
boekenvak.be/gprc

Table of Contents

About the Editors

Katleen Janssen is postdoctoral researcher at the Interdisciplinary Centre for Law and ICT of the KU Leuven. Her research focuses on legal aspects of public sector information, open government data, geographic data and spatial data infrastructures. Since 2006, she is co-chair of the INSPIRE drafting team on data and service sharing. Between 2007 and 2011, she was a member of the multidisciplinary project SPATIALIST (Spatial Data Infrastructures and Public Sector Innovation in Flanders, Belgium). In 2009 she obtained her PhD with a dissertation on the relationship between INSPIRE, the re-use of public sector information and access to environmental information.

Joep Crompvoets is senior researcher at the Public Management Institute of the KU Leuven and the Centre for Spatial Data Infrastructures and Land Administration of the University of Melbourne. Between 2007 and 2011, he was the project coordinator of the SPATIALIST project. From 2011, he is the Secretary-General of EuroSDR, a European Spatial Data Research Network, which is a not-for-profit organisation linking national mapping and cadastral agencies with research institutes and universities for the purpose of applied research in spatial data provision, management and delivery. His research interests are strongly focused on the development of spatial data infrastructures around the world.

About the Authors

Malay Adhikari is a Ph.D. research scholar in space law at the Centre for International Legal Studies of the School of International Studies at the Jawaharlal Nehru University in New Delhi. He obtained a M. Phil in space law from NALSAR University of Law in Hyderabad (India), an LL.M. degree in international law and a M.Sc. in physics. Before joining the Centre for International Legal Studies, he was a practicing lawyer.

Spiros Athanasiou is a research associate at IMIS/Athena RIC. He received his diploma in electrical engineering from NTUA, and has worked as an ICT advisor and Project/RTD manager for the public and private sector. His interests include geospatial data management, crowdsourcing services and open data. He is leading the design and development of geodata.gov.gr, the first government service in Greece that provides open data. He is the head of the Working Group for Open Data in digitalgreece2020.gr, member of OSGeo and OGC, and an active supporter of open source and open data. He was a member of the Prime Minister's eGovernment Group (egovict.blogspot.gr), with the responsibility to promote and enforce open data in the public administration.

George Cho is a professor of geoinformatics and the law at the University of Canberra and specialises in information systems, technology law and environmental issues. He is a research scientist with the Institute for Applied Ecology and the Faculty of Applied Science. His research focuses on the legal and policy issues affecting geographic information science and the practical legal problems confronting the application of technology in the creation, use and impact of digital spatial information on developers, owners and users.

Karen Donders lectures European information society at the Vrije Universiteit Brussel. She is a senior researcher with the Center for Studies on Media Information and Telecommunication (IBBT-SMIT) at the same university. Her research, funded by the Research Foundation Flanders (FWO), focuses on public service broadcasting, European competition law, commercial broadcasting, broadcasting and distribution, etc. Her monograph *Public Service Media and Policy in Europe* was published in the beginning of 2012 with Palgrave Macmillan.

Anne Fitzgerald is a professor in law research at Queensland University of Technology (QUT) Law School, Brisbane, Australia. From 2005, Anne worked on the Government Information Licensing Framework (GILF) project in collaboration with Queensland Government legal and information management experts. She was a lead legal researcher on the Cooperative Research Centre for Spatial Information project (2007-2010) that investigated the use of Creative Commons licences for public sector copyright materials. Since 2007 Anne has been the project lead for Creative Commons Australia's engagement with the Australian public sector and has overseen the adoption of CC licences as the default copyright licence for public sector materials.

Mark N. Gasson was a senior research fellow at the School of Systems Engineering, University of Reading in the UK and now holds the post of visiting research fellow. He obtained his first degree in cybernetics and control engineering in 1998 from the Department of Cybernetics at Reading, and obtained his Ph.D. in 2005 for developing an experimental invasive interface between the nervous system of a human volunteer and a computer system. His interdisciplinary research interests predominantly centre around user-centric applications of emerging technologies, with specific focus on pushing the envelope of human-machine interaction.

Eleni Kosta is a senior research fellow at the Interdisciplinary Centre for Law & ICT (ICRI) at the KU Leuven, conducting research in the field of privacy and identity management, specialising in new technologies and electronic communications. She obtained her law degree at the University of Athens, and after obtaining a Master's degree in public law at the same University, as well as an LLM in legal informatics (Rechtsinformatik) at the University of Hannover, she obtained the title of doctor of law at the Faculty of Law of the KU Leuven in Belgium. Eleni is also an associate at the Brussels based law firm time.lex and a member of the Iraklion Bar Association.

Konstantinos A. Nedas is a member of the Board of the Hellenic Mapping and Cadastral Organisation and the representative of Greece to the INSPIRE Committee of the E.U. He received a diploma in surveying engineering from the Aristotle University of Thessaloniki, Greece in 2000 and his Ph.D. from the University of Maine, USA in spatial information science and engineering in 2006. He is a member of the Greek National Geoinformation Committee. Since 2009 he is responsible for coordinating the implementation of the Greek National Geospatial Data Infrastructure, including the transposition of the INSPIRE Directive and the issuance of the first unifying licensing framework for public sector geospatial data in Greece.

Kalliope Pediaditi is a strategic environmental planner and one of the four authors of the Greek INSPIRE transposition law 3882/2010. Currently a scientific advisor at the Hellenic Mapping and Cadastral Organisation, she has also served as advisor to the Minister of Environment, Energy and Climate Change. Her Ph.D. is in sustainability metrics and she has coordinated a range of EU research projects focusing on the development of participatory evaluation, indicator and decision support systems.

Kevin Pomfret is executive director of the Centre for Spatial Law and Policy, a not-for-profit organisation that focuses on the policy and legal issues associated with the collection, use and distribution of location and other types of geospatial data. He is also a practicing attorney with a Virginia-based law firm, GeoLaw, P.C. Kevin, a former imagery analyst, speaks and writes regularly on issues of spatial law and policy and serves on the board of directors of the Open Geospatial Consortium.

Tim Raats is a researcher at the Centre for Studies on Media, Information and Telecommunication (IBBT-SMIT) at the Vrije Universiteit Brussel. He is currently working as a Ph.D. aspirant for FWO-Vlaanderen (Fund for Scientific Research), conceptualizing and assessing public broadcasting in a multi-stakeholder media ecology, setting out from a geographic-comparative analysis of public service broadcasting's cultural remit.

Prodromos Tsiavos is a solicitor in Avgerinos Law Firm in Athens and the legal project lead for the Creative Commons England and Wales and Greece projects. He is a legal adviser for the National Documentation Center/National Hellenic Research Foundation and the Hellenic Association of Academic Libraries. He is a lecturer on eGovernment and open data management in the Greek National School of Public Administration. He read law in Athens and London and holds a Ph.D. in information systems from LSE. Prodromos was part of the legal team that drafted Law 3882/2010 transposing the INSPIRE Directive into Greek Law.

Niels Vandezande is a legal researcher at the Interdisciplinary Centre for Law & ICT, Faculty of Law of the KU Leuven since September 2009. He obtained his master of laws degree in 2009 at the same university, focusing on ICT-law, international law and corporate law. He currently conducts research in the fields of data protection, privacy, security & trust and public information availability.

Bastiaan van Loenen is assistant professor at Delft University of Technology. His research focuses on the legal and administrative aspects evolving around the collection, processing and use of geographic information. Currently his main focus is the stimulation of re-use of public sector geo-information. Bastiaan chairs the legal and socio-economic committee of the Global Spatial Data Infrastructure Association (GSDI), is a member of the Geo-information infrastructure committee of the Dutch Geodetic Commission, and editor of GIM international. In 2010, van Loenen was awarded the professor J.M. Tienstra prize for most talented young Dutch researcher in the geo-sciences domain.

Frederika Welle Donker was born in the Netherlands, and emigrated to Sydney, Australia, after leaving high school. In Sydney she worked as a technical officer at the Commonwealth Scientific & Industrial Research Organisation and at the University of Sydney. She returned to The Netherlands in 1998 and completed her M.Sc. at the Faculty of Technology, Policy & Management of Delft University of Technology in 2001. Since 2005 Frederika has been a researcher at the OTB Research Institute for the Built Environment, Delft University of Technology. In 2009 she started a Ph.D. study into accessibility aspects of public sector information for re-users and expects to complete her thesis in 2012.

List of Abbreviations

ABS	Australian Bureau of Statistics
ANZLIC	Australia New Zealand Land Information Council
AODCJF	Australian Ocean Data Centre Joint Facility
AODN	Australian Ocean Data Network
API	Application Program Interface
APIE	Agence du Patrimoine Immatériel de l'État (Agency for Public Intangibles of France)
AUSGOAL	Australian Government Open Access and Licensing framework
AWRIS	Australian Water Resources Information System
BBC	British Broadcasting Corporation
BOM	Bureau of Meteorology
BY	Attribution
CC	Creative Commons
CC0	Creative Commons Zero
CCTV	Closed Circuit Television
DG	Directorate General
DRM	Digital Rights Management
EC	European Community
ECJ	Court of Justice of the European Union
ECOMET	Economic Interest Grouping of National Meteorological Services of the European Union
ERMI	Electronic Rights Management Information
ERTS	Earth Resource Technology Satellite
ESDIN	European Spatial Data Infrastructure Network
EU	European Union
FEMA	Federal Emergency Management Agency
FOI	Freedom of Information
FOIA	Freedom of Information Act
FTC	Federal Trade Commission
GA	Geoscience Australia
GATS	General Agreement on Trade in Services
GEOSS	Global Earth Observation Systems of Systems
GI	Geographic Information
GILF	Geographic Information Licensing Framework
GIPA	Government Information Public Access
GIS	Geographic Information System
GMES	Global Monitoring for Environment and Security
GMS	Gaelic Media Service
GPS	Global Positioning System
GSM	Global System for Mobile communication
GSDI	Global Spatial Data Infrastructure
HEMCO	Hellenic Mapping and Cadastral Organisation
HR	High Resolution
INSPIRE	Infrastructure for Spatial Information in Europe
IP	Intellectual Property
IPR	Intellectual Property Rights

ISRO	Indian Space Research Organisation
IWT	Agentschap voor Innovatie door Wetenschap en Technologie (Agency for Innovation by Science and Technology)
JSG	Joint Steering Group
KNMI	Koninklijk Nederlands Meteorologisch Instituut (Royal Dutch Meteorological Institute)
MEECC	Ministry of Environment, Energy and Climate Change
NC	Non-Commercial
ND	No Derivatives
NGC	National Geo-information Committee
NIFGIS	National Interoperability Framework on Geographic Information and Services
NPG	National Policy on Geo-information
NRSC	National Remote Sensing Centre
NSDI	National Spatial Data Infrastructure
OECD	Organisation for Economic Co-operation and Development
OFT	Office of Fair Trading
OMB	Office of Management and Budget
OST	Outer Space Treaty
PDM	Public Domain Mark
POI	Point of Interest
PSI	Public Sector Information
RFID	Radio Frequency Identification
RSDP	Remote Sensing Data Policy
RTI	Right to Information
SA	Share Alike
SCA	Stored Communications Act
SDI	Spatial Data Infrastructure
SHOM	Service Hydrographique et Océanographique de la Marine (Marine Hydrological and Oceanographic Service)
SMS	Short Message Services
TFEU	Treaty on the Functioning of the European Union
TIROS	Television Infrared Observation Satellite
TPM	Technological Protection Measures
UMTS	Universal Mobile Telecommunications Systems
UN	United Nations
VGI	Volunteered Geographic Information
VHS	Video Home System
WTO	World Trade Organisation

INTRODUCTION

Introduction

Katleen Janssen[1] *and Joep Crompvoets*[2]

Setting the scene

The use of geographic data is becoming increasingly commonplace. Satellite navigation systems and smartphones with digital maps are part of almost every western household, and many popular websites, services and applications (apps) are based on geographic data. Obvious examples are Google Maps and Google Street View, but many other applications can be thought of, such as weather information, geo-caching, road maintenance, emergency response to flooding or earthquakes, or fertilisation optimisation.

The growing importance of geographic data for our society becomes clear from the many local, national, European and international initiatives taken to exchange these data between governments, and to make them available to citizens, organisations and companies. For example, in the European Union, the INSPIRE initiative set up an infrastructure for the sharing of geographic data[3], and globally the United Nations and the Group on Earth Observations are working on facilitating the exchange of earth observation data[4].

Geographic data

Geographic data identify or describe locations on the surface of the Earth. They can include e.g.addresses, market research data, census data, health data, data on environment and natural resources, descriptions of transportation and utility networks, information on flows of goods, cadastral and land registries, or data obtained by remote sensing from satellites.

In the past such information took the form of paper maps. Maps have been preserved on Babylonian clay tablets from about 2300 BC[5]. Modern mapping dates from the Age of Discovery in the fifteenth and sixteenth centuries, a time when trade, travel and migration increased exponentially. As geographic knowledge was also of vital importance in military campaigns, many of the national mapping agencies were established under the auspices of the military, such as the Survey of India in 1767 and the Ordnance Survey of Great Britain in 1791.

With the development of information technology, also the representation of geographic data evolved. Maps were digitised and spatial databases were created in which data are stored in a logical manner. Geographic Information Systems (GIS) were developed: computer systems for capturing, managing, integrating, manipulating, analysing and displaying data that are spatially referenced to the Earth. They were first developed during the 1960s, but it was not until the mid 1980s that computers became sufficiently performant to handle the considerable amounts of data in these systems. In 1987 the British Government Committee of Enquiry

---

1 Interdisciplinary Centre for Law and ICT, KU Leuven, katleen.janssen@law.kuleuven.be.
2 Public Management Institute, KU Leuven, joep.crompvoets@soc.kuleuven.be.
3 See http://inspire.jrc.ec.europa.eu/ (accessed on 18/10/2011).
4 See http://www.earthobservations.org/index.shtml (accessed on 18/10/2011).
5 I. Masser (2010). *Building European Spatial Data Infrastructures*. Redlands: ESRI Press.

chaired by Lord Chorley hailed the advent of GIS as 'the biggest step forward in the handling of geographic data since the invention of the map'[6]. Developments since have massively increased computer storage and handling capacities, and the Internet has made it possible to transfer large amounts of geographic data between machines, people and organisations.

This possibility, combined with the growing need of public bodies, companies and organisations to share geographic data between institutions and across borders, has led to the movement from GIS to spatial data infrastructures (SDI) from the early 1990s onward. An SDI is 'the means to assemble geographic data that describes the arrangement and attributes of features and phenomena on the Earth. The infrastructure includes the materials, technology, and people necessary to acquire, process, and distribute such data to meet a wide variety of needs'[7]. The overriding objective of SDIs is to facilitate access to geographic data held by a wide range of stakeholders in both the public and private sectors with a view to maximising overall usage and supporting decision-making. An SDI includes not only the technology for exchanging data (software, standards, web services), but also non-technical components such as legislation and policies, know-how, funding, coordination, human resources and partnerships[8].

Producers and users of geographic data

A large part of the current geographic data is created by government and the public sector in the course of fulfilling their public tasks, e.g. cadastres, national mapping agencies, statistic offices, and meteorological services, but also ministries, counties and municipalities. They need these data in order to prepare, implement and evaluate their policies and to provide efficient and effective services to companies and citizens, for instance in the domains of urban and regional planning, transportation, housing, economic development, emergency management, security, or property tax collection.

Geographic data are also of vital importance for the private sector. In a constantly changing marketplace, successful businesses must strive to reduce costs, improve profit margins, and maintain their competitive edge to retain customers. Geographic data enable businesses to adapt their business strategies to market developments. For instance, the real estate industry has a strong interest in geographic data on individual properties and the general land market for financial investment and property development purposes. Retail companies also need geographic data to complement their customer databases and help them find the best locations for new stores and offices. In addition, many private sector companies operating in different domains are adding value to existing geographic data to sell new information products and services on the market, such as navigation systems, weather services, environmental reports, or tourism services. Generally, the producers of these value-added products and services look towards public sector geographic data as a resource. However, several private sector companies also produce large geographic datasets,

[6] Department of the Environment (1987). *Handling geographic information. Report of the Committee of Enquiry chaired by Lord Chorley*. London: HMSO.

[7] National Research Council (1993). *Toward a coordinated spatial data infrastructure for the nation*. Washington: National Academies Press.

[8] L. Grus, J. Crompvoets and A. Bregt (2007). "Multi-view SDI assessment Framework". *International Journal of Spatial Data Infrastructures Research* 2, 33-53.

such as Teleatlas/TomTom and Navteq/Nokia, and private sector aerial photography or satellite imagery providers.

Not only companies, but also citizens benefit from the availability of geographic data. They use geographic data to check the weather, to organise their route to school or work, or to examine whether their property lies in a polluted area. While until recently citizens could be considered as passive receivers of geographic data created by the public sector and the information industry, the development of new technologies and the wider availability of broadband access have enabled them to create their own geographic data. This development is part of the general increase in user-generated content and the development of 'Web 2.0'[9]. With regard to geographic data, the term 'Volunteered Geographic Information' (VGI) or 'Neogeography' refers to the widespread creation of digital geographic data by citizens on a voluntary basis and generally without any formal or professional qualifications[10]. Examples of such VGI include OpenStreetMap[11], Foursquare[12] and Wikimapia[13].

Need for legal research

The growing importance of geographic data and SDIs in our society, combined with the increased interaction between different stakeholders, increases the risks for disputes and the need for regulation of the stakeholders' behaviour in the SDI. However, while the focus on technical issues of geographic data is currently being replaced by more attention for the non-technical, organisational and 'human' aspects of an SDI, these aspects are still under development. In many cases, they take more time and effort to establish than the technical aspects, because they require changes in attitude and culture, which are part of a long-term process. While the non-technical knowledge with regard to geographic data and SDI has been growing[14], the research and practice on legal aspects has still not reached a sufficient level of maturity.

A number of factors play a role in this. First, legal solutions are traditionally sought on a national level, while the need for geographic data and the development of SDIs no longer stop at national borders. This requires a change in the mindset of the lawmakers and the legal scholars and practitioners, as they have to deal with legal questions and concerns at a new cross-border, international, or even global level. Second, it is common that the development of the law lags behind on changes in society, economy, technology, and culture. Law only reacts to phenomena occurring in reality, rather than steering these phenomena from

[9] T. O'Reilly (2007). "What is Web 2.0: Design Patterns and Business Models for the Next Generation of Software". *Communications and Strategies* 65, 17-37.

[10] M. Goodchild (2009). "Neography and the nature of geographic expertise". *Journal of Location Based Services* 3(2), 82-96; S. Elwood (2008). "Volunteered geographic information: future research directions motivated by critical, participatory, and feminist GIS". *GeoJournal* 72(3-4), 173-183; D. Coleman, Y. Georgiadou and J. Lebonte (2009). "Volunteered Geographic Information: the nature and motivation of produsers". *International Journal of Spatial Data Infrastructures Research* 4, 332-358.

[11] See http://www.openstreetmap.org/ (accessed on 18/10/2011).

[12] See https://foursquare.com/ (accessed on 18/10/2011).

[13] See http://wikimapia.org/ (accessed on 18/10/2011).

[14] Z. Nedović-Budić, J. Crompvoets and Y. Georgiadou (ed.) (2011). *Spatial Data Infrastructures in Context: North and South*. Boca Raton: CRC-Press; J. Crompvoets et al. (2008). "Interdisciplinary Research Project: SPATIALIST. Spatial Data Infrastructures and Public Sector Innovation in Flanders (Belgium)". *Proceedings of GSDI-10 Conference, Small Island Perspectives on Global Challenges: The Role of Spatial Data in Supporting a Sustainable Future*. St. Augustine, Trinidad, St.Augustine: GSDI.

the start. However, in order to avoid the usual time gap between the emergence of new technologies or new societal preferences and the development of regulations to keep up with these changes, the legal questions should be discussed earlier in the process, together with the technical, economic, and organisational matters[15].

This book aims to contribute to the legal research and practice on geographic data and SDIs by addressing some of the main challenges we are currently facing. The different contributions in the book show the legal questions and difficulties that are arising and that urgently need to be answered, for instance with regard to privacy, liability and licensing. On the other hand, they also provide some examples on new research trying to tackle these challenges, or showcase some good practices on how law and policy are already being used in a more proactive way to influence the development of SDIs.

Regulatory framework for geographic data and SDIs

In order to assess the new legal challenges that arise in the creation and exchange of geographic data and the development of SDIs, one needs to have an insight in the scope and composition of the applicable legal framework. This legal framework consists of a broad set of rules and regulations, aiming to organise a particular element in society (in this case the exchange of geographic data in an SDI). These rules and regulations were not necessarily developed specifically for this particular object, but may have been created for other domains in society and are now also applied to geographic data and SDIs. This can include legislation that deals with information, data or content, such as freedom of information, intellectual property rights or the protection of personal data. It can also involve legislation and policy with an even broader scope, such as tort liability and contract law, which apply to any kind of actor, situation or object falling within the field of application.

The regulatory framework for geographic data and SDI consists of three types of laws, regulations and policies[16]. These categories are not always in agreement and sometimes contradict each other. The first category includes legislation and policies that stimulate the availability of geographic data within the SDI to all stakeholders. Within this first category, three subcategories can be distinguished. First, particular laws organise the exchange or *sharing* of data between public bodies for the purpose of policymaking or service delivery to the citizens, i.e. for the performance of public services or public tasks. This includes e.g. the European Union's INSPIRE Directive[17]. Second, there is an extensive body of legislation that provides *access* to geographic data for the citizens, in order to increase transparency, government accountability and public participation. Examples of such legislation include national Freedom of Information Acts; the 1998 Aarhus Convention on access to information,

[15] K. Janssen and J. Dumortier (2007). "Legal Framework for a European Union Spatial Data Infrastructure: Uncrossing the Wires". In Onsrud, H. (ed.). *Research and Theory in Advancing Spatial Data Infrastructure Concepts.* Redlands: ESRI Press, 231-244.

[16] K. Janssen (2010). *The Availability of Spatial and Environmental Data in the European Union. At the Crossroads between Public and Economic Interests.* Alphen a/d Rijn: Kluwer Law International; K. Janssen (2008). "A legal approach to assessing Spatial Data Infrastructures". In Crompvoets, J. et al. (ed.). *A Multi-View Framework to Assess Spatial Data Infrastructures.* Melbourne: University of Melbourne, 255-272.

[17] European Parliament and Council (2007). Directive 2007/2/EC of 14 March 2007 establishing an Infrastructure for Spatial Information in the European Community (INSPIRE), *OJ L* 108, 1.

public participation in decision-making and access to justice in environmental matters[18]; and the 2003 European Directive on public access to environmental information.[19] Third, the economic interests in the availability of geographic data are represented by legislation or policies addressing the use or *re-use* of geographic data for economic purposes, such as the creation of added value information products or services. The most well-known example of this type of rules is the 2003 European Directive on the re-use of public sector information.[20]

The second category of laws, regulations and policies has the intention or effect of restricting the availability of geographic data in the SDI. These protect legitimate interests and entail that geographic data may not always be freely available for any use. At least four types of interests can be considered. First, the availability of geographic data may sometimes cause concerns for the privacy of the individual, as these data may allow others (e.g. government, law enforcement, commercial companies) to learn the whereabouts of an individual, his behaviour or his relationship to a particular place or area. In this way, geographic data can provide extensive information about one's private life. Legislation protecting privacy includes the European Convention protecting Human Rights and Fundamental Freedoms[21], the United States Fourth Amendment, or the 1995 European Directive on the processing of personal data[22]. A second interest that may lead to a limited availability of geographic data is national security. The security of a State or nation may be put at risk by the publication of data on the location of e.g. critical infrastructure. An example of policies restricting the availability of geographic data for security reasons is the limited availability of remote sensing data under the Indian Remote Sensing Data Policy. Third, liability provisions may bring geographic data providers to strictly controlling what happens with their data. Particular geographic data may have been created for a certain purpose, and its use for other objectives may lead to errors or accidents and subsequently to damage to persons or goods. The provider of geographic data may be held responsible for the damage, or may have to fulfil certain conditions when making his data available. Applicable legislation includes national rules on tort and contract liability, or the 1985 European Directive on product liability[23]. Finally, the fourth type of regulations that restrict the availability of geographic data is the legislation on the protection of intellectual property rights (IPR). The objective of IPR is to stimulate innovation and creativity by rewarding authors for their work with an exclusive right to decide on its publication or further use. Provisions protecting IPR can for instance be found in the 1886 Berne Convention for the Protection of Literary and

---

[18] United Nations Economic Committee for Europe. (1998). Convention on access to information, public participation in decision-making and access to justice in environmental matters, http://www.unece.org/env/pp/documents/cep43e.pdf (accessed on 24/10/2011).

[19] European Parliament and Council (2003). Directive 2003/4/EC of 28 January 2003 on public access to environmental information and repealing Council Directive 90/313/EEC, *OJ L* 41, 26.

[20] European Parliament and Council (2003). Directive 2003/98/EC of 17 November 2003 on the re-use of public sector information, *OJ L* 245, 90.

[21] Council of Europe (1950). Convention for the Protection of Human Rights and Fundamental Freedoms, http://conventions.coe.int/Treaty/en/Treaties/html/005.htm (accessed on 24/10/2011).

[22] European Parliament and Council of the European Union (1995). Directive 95/46/EC of 24 October 1995 on the processing on the protection of individuals with regard to the processing of personal data and on the free movement of such data, *OJ L* 281, 31.

[23] Council (1985). Directive on the approximation of the laws, regulations and administrative provisions of the Member States concerning liability for defective products, *OJ L* 210, 29.

Artistic Works[24], the 1996 European Directive on the legal protection of databases[25], and national IPR legislation.

The third type of laws, regulations and policies that are part of the legal framework for geographic data and SDI deals with the underlying relationships between the stakeholders. Two main types can be considered here. First, the conflicting relationship between private competitors or between the public and the private sector on the market is dealt with by competition law, antitrust and rules on the maintenance of a level playing field. Second, the more cooperative relationship between the stakeholders is dealt with by rules on coordinating bodies, public-private partnerships, citizen representation, procedures for cooperation agreements, etc.

The legal framework for geographic data and SDI is depicted in a simplified manner in Figure 1.

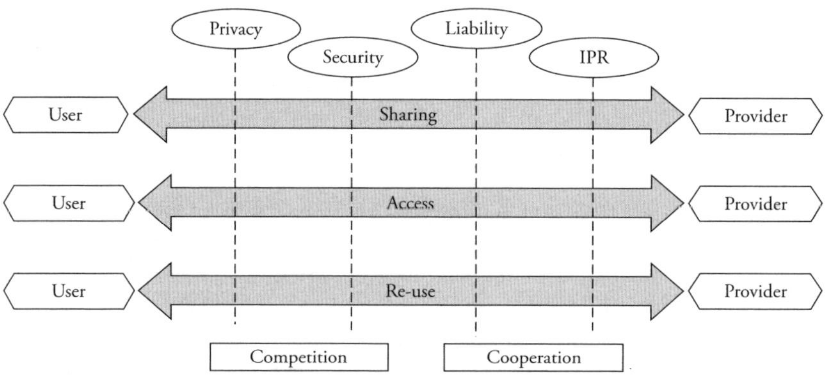

Figure 1: Legal framework for geographic data[26]

The contributions in this book

This book includes nine contributions that illustrate clearly the main legal aspects regarding the availability of geographic data. These contributions are divided into three parts: 1) policies promoting availability, 2) policies restricting availability, and 3) the relationship between the public and private sector. Based on these contributions, the conclusions of this book aim to indicate the challenges and priorities for future research and policy.

The contributions are based on current regulation and practice influencing the availability or the protection of geographic data, include a critical analysis of the strengths and weaknesses of these regulations and practice, and assess their impact on the data availability for different stakeholders. The authors are scholars and practitioners from different fields of expertise, different continents and different legal traditions. In this way,

[24] Berne Convention (1886). Convention of 9 September 1886 for the Protection of Literary and Artistic Works, http://www.wipo.int/treaties/en/ip/berne/ (accessed on 24/10/2011).

[25] European Parliament and Council of the European Union (1996). Directive 96/9/EC of 11 March 1996 on the legal protection of databases, *OJ L* 77, 20.

[26] Based on K. Janssen (2010). *The Availability of Spatial and Environmental Data in the European Union. At the Crossroads between Public and Economic Interests.* Alphen a/d Rijn: Kluwer Law International.

one-sided views based on an EU-centric approach to the availability of geographic data are avoided.

In Part 1, three contributions deal with legal issues that promote the availability of geographic data. The first contribution of Bastiaan van Loenen, Katleen Janssen and Frederika Welle Donker, entitled *Towards true interoperable geographic data: developing a global standard for geo-data licenses,* deals with the strong need to harmonise licensing models. Based on an examination of the commonalities and differences between existing licensing models, it explores the possibility of developing a global licensing model that could increase the transparency of conditions for accessing and using geographic data.

The second contribution of Prodromos Tsiavos, Kalliope Pediaditi, Costas Nedas and Spyros Athanasiou, *Cultivating open geodata ecologies: Lessons from the implementation of Directive 2007/2/EC,* discusses the Greek law 3882/2010 entitled 'National Infrastructure for Geospatial Information'. This law constitutes the Greek national transposition of the INSPIRE Directive. It adopts a data-life cycle approach and increases the possibilities for re-use of public sector information. In that sense, it could operate as a model for a general system for organizing the regulation of open data in the public sector. The authors argue that it is important to view the transposition of the INSPIRE Directive as part of a greater ecology of laws aiming at the cultivation of a broader common information environment. Such ecology contains legislative instruments, but also ministerial decrees, licensing practices and interoperability specifications.

The third contribution, written by Anne Fitzgerald, deals with *Juggling information policy, rights to information and copyright licensing to enhance the accessibility and reusability of spatial data in Australia.* This contribution describes the significant developments in information policy and practice in Australia in the last five years, which have resulted in a considerable lessening of the barriers to accessibility and reusability of geographic data and other material held by public sector agencies. The Government Information Licensing Framework provides access to public sector information under an open content copyright-based regime, applying the Creative Commons suite of licenses to standardise the conditions for use and re-use of licensed materials. This approach is based on the recognition that re-use of public sector materials can be facilitated by directly addressing the obstacles presented by the complex, protracted and restrictive licensing practices and procedures that are typically encountered.

In Part 2, four contributions deal with legal issues that restrict the availability of geographic data. The first contribution by Kevin Pomfret, *Summary of location privacy in the United States,* argues that the existing legal and policy framework in the United States is unclear with respect to what is privacy from a location standpoint. Legislation is outdated and there is very little in terms of applicable case law. As a result, policymakers, businesses and the general public are all struggling to define 'a reasonable expectation of privacy' from a location standpoint.

The second contribution of Part 2, written by Eleni Kosta and Mark Gasson, is entitled *The processing of location data under the European data protection legal framework: behavioural profiling via location tracking.* This contribution focuses on an experiment to explore the information that can be revealed about a user from the collection and processing of his or her location data, and assesses the potential impact on the privacy of the user in the light of the European data protection legislation. In addition to the examination of the current European legal framework on the processing of location data, the implications of this framework for the deployment of location based services are discussed.

The third contribution, *Geographic data and legal liability issues*, by George Cho, addresses liability questions relating to the use of geographic information systems. It suggests that conventional legal solutions may have a lesser role to play in the 21st century as new liability issues emanate from different sources and present themselves in a variety of forms and degrees of complexity. Cho demonstrates this through the examples of global positioning systems and the Web 2.0 environment.

In the final contribution, *High Resolution Data Dissemination in the Background of National Security of India*, Malay Adhikari discusses the conflict between the restriction of high resolution satellite imagery resources for national security reasons and the promotion of their use for commercial reasons. He argues that while India may have a problem with its national security, it should not completely ignore the potential and opportunities of high resolution data. It is recommended to keep the market of high resolution open, yet being aware that there must be some specific and strict regulations to punish any infringement on national security.

The two contributions of Part 3 are about regulations and policies dealing with the underlying relationships between different stakeholders groups, more specifically the relationship between the public and private sector. The first contribution, by Katleen Janssen and Niels Vandezande, is entitled *The relationship between the public and private sector in the spatial information market: how to define their roles?* This contribution addresses the debate on the role of the public sector on the European market for geographic information services. It explains that this debate has been given a new dimension by the introduction of the European Directive on the re-use of public sector information. In addition, Janssen and Vandezande focus on the possible concerns and barriers raised by the activities of the public sector on the geographic information market, against the background of European competition law. Based on this analysis, they address some criteria for assessing whether the public sector should operate on the information market.

The second contribution, *Measuring public value with the Public Value Test: best or worst practice?* is from Karen Donders and Tim Raats. This contribution provides some insights in the relationship between the public and private sector from an outside perspective. The geographic information market is not the only one that has to deal with this debate. Another domain that has been struggling with the division of roles between the public and private sector is public broadcasting. Donders and Raats examine how the relationship between the public broadcaster and its private counterparts is being tackled by using the example of the United Kingdom's Public Value Test.

The book ends with some conclusions on future research challenges and priorities that can be drawn from the different contributions. It argues that these challenges need to be addressed by legal researchers in a multidisciplinary environment, taking into account technical, institutional and cultural factors influencing the availability of geographic data and the functioning of SDIs.

Note: the information in this book has been updated until August 2011.

Policies Promoting Availability

Part 1

Towards True Interoperable Geographic Data: Developing a Global Standard for Geo-data Licences

Bastiaan van Loenen[1], *Katleen Janssen*[2] *and Frederika Welle Donker*[3]

Introduction

In the development and diffusion of major technologies and applications, standards provide an essential input, establishing compatibility or interoperability between components in a system[4]. This is also true for the case of the dissemination of geographic data. Not only technical standards are important, but also agreements establishing interoperability on an organisational or legal level. However, while the harmonisation of licensing models and practices is essential for sharing geographic data across borders between public bodies, the private sector and broader society, the focus in the geo-domain still remains mostly on the standardisation of the technical aspects of geographic data. Yet, non-transparent and inconsistent licences have often been identified as a major barrier to the sharing of data across the geospatial community and a clear need for harmonised geo-licences is increasingly being recognised[5].

Currently, it is very difficult to readily assess and directly access geographic data and geographic information services. A major barrier is the use of non-standard licences that are difficult to understand, both for human beings and computers. This applies to the use and combination of different geographic data and geographic information services within one jurisdiction, but also to cross-border and international use. For example, INSPIRE will promote the technical interoperability of geographic data allowing different data sets across Europe to be smoothly combined in new data sets and/or services[6]. However, without

[1] Delft University of Technology, B.vanLoenen@tudelft.nl.

[2] Interdisciplinary Centre for Law and ICT, KU Leuven, katleen.janssen@law.kuleuven.be.

[3] Delft University of Technology, F.M.WelleDonker@tudelft.nl.

[4] Department of Trade and Industry (2005). *The empirical economics of standards*, www.bis.gov.uk/files/file9655.pdf (accessed on 26/08/2011).

[5] MICUS (2008). *Assessment of the re-use of public sector information (PSI) in the geographic information, meteorological information and legal information sector. Final report*, http://ec.europa.eu/information_society/policy/psi/docs/pdfs/micus_report_december2008.pdf (accessed on 26/08/2011); N. Groot et al. (2007). "Toegankelijkheid publieke geodata vooral gehinderd door juridische barrieres". *VIMatrix* 15(8), 6-8 [in Dutch]; B. van Loenen et al. (2007). *Open toegankelijkheidsbeleid voor geo-informatie vergeleken: het gras leek groener dan het was*. Delft: Delft University of Technology [in Dutch]; National Research Council (2004). *Licensing Geographic Data and Services*. Washington: National Academies Press; Spatial Technologies Industry Association (2001). *Phase 1 report: increase private sector awareness of, and enthusiastic participation in the National Spatial Data Infrastructure (NSDI)*, http://www.fgdc.gov/library/whitepapers-reports/sponsored-reports/stia/ (accessed on 26/08/2011); RAVI Bedrijvenplatform (2000). *Economische effecten van laagdrempelige beschikkingstelling van overheidsinformatie*. Amersfoort: RAVI Bedrijvenplatform [in Dutch]; H. Meixner and A.U. Frank (1997). *Study on Policy Issues Relating to Geographic Information in Europe*, ftp://ftp.geoinfo.tuwien.ac.at/frank/2184_GIPolicy.pdf (accessed on 26/08/2011).

[6] See European Parliament and Council (2007). Directive 2007/2/EC of 14 March 2007 establishing an Infrastructure for Spatial Information in the European Community (INSPIRE), *OJ L* 108, 1.

standardised licences these new data sets and/or services will either never be developed or be developed with significant delay.

Only recently the development towards a standard for geo-information licences has started to emerge. Examples of promising initiatives aimed at standardising licences are e.g. found in the United States[7], Europe[8], Italy[9], the Netherlands[10], Australia[11], and at a global level[12]. However, there are increasing calls for licensing models that have a broader reach than just on a national or sector level, possibly based on existing models such as Creative Commons or Open Data Commons[13].

During its meeting at the GSDI 12 conference in Singapore in October 2010, the Legal and Socio-Economic Working Group of the Global Spatial Data Infrastructure Association (GSDI) felt that the possibilities for such global licensing models for geographic data needed to be examined further. The Group believes that the differences between the national traditions and practices with regard to licensing might actually be smaller than generally assumed, entailing that efforts to harmonise these traditions and practices may be worthwhile. The task of looking for harmonisation of licence models was taken up by a number of academics and practitioners, who aim to take some first steps towards a global approach to licensing.

First, the Working Group collected existing material on (national and international) licensing frameworks, compared the key components thereof and categorized them in a number of 'common denominator' groups. The next step is to develop a framework of several types of licences based on these categories, which can be used globally and will increase transparency of the conditions for obtaining and using geographic data. The intention of the framework is to avoid creating new licensing conditions if this is not necessary, but also to accommodate possible differences between organisations, cultures, and financing models. This paper presents the first stage of the work, i.e. the comparison between several existing licensing models and the definition of common elements that can serve as a basis for a global set of model licences.

---

[7] National Research Council (2004). *Licensing Geographic Data and Services*. Washington: National Academies Press.

[8] INSPIRE Drafting Team Data and Service Sharing (2010). *Guidance on the 'Regulation on access to spatial data sets and services of the Member States by Community institutions and bodies under harmonised conditions*, http://inspire.jrc.ec.europa.eu/documents/Data_and_Service_Sharing/DSSDraftGuidancedocument_v4.1.pdf (accessed on 26/08/2011).

[9] L. Garretti et al. (2009). "Standard Licences for Geographic Information: the Development and Implementation in Local Government in Italy". In van Loenen, B., Zevenbergen, J.A. and Besemer, J.W.J. (ed.). *SDI Convergence - Research, Emerging Trends, and Critical Assessment*. Delft: NCG, 53-62.

[10] F. Welle Donker, B. van Loenen and J. Zevenbergen (2010). "Geo Shared licences: a base for better access to public sector geoinformation for value-added resellers in Europe". *Environment and Planning B: Planning and Design* 37, 326-343.

[11] A. Fitzgerald (2010). "Legal Issues and Experiences from an Australia Perspective". Presentation at GSDI-12 Conference 'Realising Spatially Enabled Societies'. Singapore, 19-22 October.

[12] H. Onsrud, J. Campbell, and B. van Loenen (2010). "Interoperable Open Access Licenses for the Global Earth Observation System of Systems (GEOSS)". *International Journal of Spatial Data Infrastructures Research* 5, 194-215.

[13] See Group on Earth Observations (2010). *GEOSS Data Sharing Action Plan*, http://www.earthobservations.org/documents/geo_vii/07_GEOSS%20Data%20Sharing%20Action%20Plan%20Rev2.pdf (accessed on 18/02/2011); European Commission (2011). *Overview of the responses to the consultation on Directive 2003/98/EC on the re-use of public sector information*, http://ec.europa.eu/information_society/policy/psi/docs/consultations/cons2010/responses.xls (accessed on 18/02/2011).

Why standards?

Standards facilitate coordination and reduce complexity[14]. Historical studies of technological development (e.g. the triumph of QWERTY keyboards – to avoid letter hammers 'hanging' in type writers, and still in use for computers[15] – and the struggle between VHS and Betamax) show that standards sometimes evolve gradually in the interplay between companies, public organisations and 'standardisers' on the one hand and more market-driven processes on the other hand[16]. The advantages and drawbacks of standards can be summarized as follows.

Table 1: Pros and cons of standards[17]

Pro	Con
• represent an effective way of transmitting information; • facilitate co-ordination and communication (especially where there is no central authority); • promote compatibility; • reduce complexity & result in simplicity; • bring advantages of large scale production; • provide optimal or at least good solutions, policies, products; • facilitate potential market development and control issues; • increase production and distribution cost efficiencies; • improve potential efficiency in use of products; • allow adapting to a world of increasing interdependence; • provide a comparative advantage (enable to compare); • increase convenience.	• benefits of co-ordination, communication and simplification are also provided by rules in general; • goals can also be achieved – often more effectively – by market processes; • can be a threat to the individual actor's freedom and right to be different and to innovate; • may impede innovation in general; • put power in the hands of large companies and experts; • drafting process is undemocratic compared to formal legislation; • may lead to poorer solutions and/or leave us with legacy (e.g. qwerty keyboards); • encourage competition for supremacy between 'standardisers'.

In many fields of technology, one can note a general process of standardisation, moving from organisational standards to global standards via national and/or international standardisation[18]. This is demonstrated in Figure 1.

[14] N. Brunsson et al. (2000). *A World of Standards*. Oxford: Oxford University Press; K. Krechmer (2009). "Open standards: A Call for Change". *IEEE Communications Magazine (May)*, 88-94.

[15] P. A. David (1985). "Clio and the economics of QWERTY". *American Economic Review* 75(2), 332-336.

[16] T.P. Hughes (1987). "The evolution of large technological systems". In Bijker, W. et al. (ed.). *The Social Construction of Technological Systems*. Cambridge: MIT Press, 51-82; R. Werle (1998). "An Institutional Approach to Technology". *Science Studies* 11(1), 3-18.

[17] Based on N. Brunsson et al. (2000). *A World of Standards*. Oxford: Oxford University Press.

[18] An example may be the electricity plug. First, every device had its own type of plug and power point. Now, national standards are agreed upon, but an international standard is still lacking. The gap between the different national standards is filled by the private sector, which provides adaptor plugs for frequent travellers. Similar development is seen with the battery charger for mobile phones. Each mobile phone had its own type of charger, not exchangeable with other (types of) mobile phones. Recently the European Union has agreed upon requiring one common charger for all electronic devices.

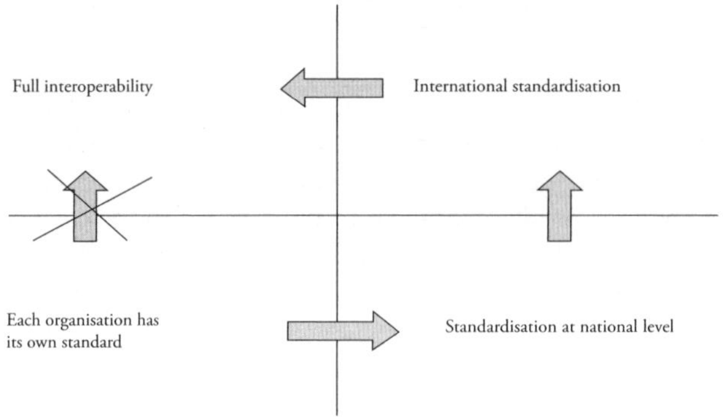

Figure 1: Suggested stages of development of standards

Similar to these standardisation processes, a standard for geo-licences may also emerge. The 'one world' principle of Krechmer might be taken as a leading principle towards further harmonisation of geographic data licences[19]. This is the principle of a single world-wide standard for a single purpose, for example to prevent technical barriers to trade. It implies that there is no need for a national or regional standard if a global standard suffices.

Start of licensing standards: Creative Commons

It can be argued that the standardisation of licences at a global scale started with the foundation of the Creative Commons organisation in 2001[20]. Many licensing initiatives in the geo-sector build on the Creative Commons licensing framework. In this section, we describe the Creative Commons framework, and discuss some of its advantages and drawbacks with regard to the harmonisation of geo-licences.

The Creative Commons framework

Creative Commons (CC) was founded as a non-profit organisation and aims to offer flexible copyright licences for creative works such as text articles, music and graphics[21]. It advocates a system whereby right holders can make works available through the Internet without forfeiting their intellectual property rights (IPR). To facilitate this, CC has developed a system of so-called CC licences, which try to balance between the 'all rights reserved' concept of traditional IPR holders and the 'no rights reserved' concept of the Public Domain, by employing a 'some rights reserved' approach[22].

---

[19] K. Krechmer (2009). "Open standards: A Call for Change". *IEEE Communications Magazine (May)*, 88-94.

[20] M. Dulong de Rosnay (2010), *Creative Commons Licenses Legal Pitfalls: Incompatibilities and Solutions*, http://www. creativecommons.nl/downloads/101220cc_incompatibilityfinal.pdf (accessed on 26/08/2011).

[21] See http://creativecommons.org (accessed on 26/08/2011).

[22] See also S. Dusollier (2006). "The Master's Tools v. The Master's House: Creative Commons v. Copyright". *Columbia Journal of Law & Arts* 29, 271-293; M. Dulong de Rosnay (2010). *Creative Commons Licenses Legal Pitfalls: Incompatibilities and Solutions*, http://www.creativecommons.nl/downloads/101220cc_incompatibilityfinal.pdf (accessed on 26/08/2011).

CC licences are model licences that offer the rightholder a limited choice in imposing certain requirements or allowing certain types of use, with regard to copying and redistribution, commercial and non-commercial use, the creation of derivative products, and the extension of the same licence conditions to derivative products. Apart from a number of standard clauses[23] and the requirement for attribution, the six CC licences offer a choice for the right holder to impose one or more of the following terms:

- The licensee may or may not use the work for commercial purposes;
- The licensee may or may not use the work to make derivative products, and if yes then:
- The licensee is or is not obliged to make the derivative product available to third parties under the same licence agreement terms.

The possible restrictions on the use that can be made of the work are summarised in the table below.

Table 2: Creative commons attributes

Symbol		Layman text
BY	Attribution	You let others copy, distribute, display, and perform your copyrighted work – and derivative works based upon it – but only if they give credit the way you request.
SA	Share Alike	You let others distribute derivative works only under a licence identical to the licence that governs your work.
NC	Non-Commercial	You let others copy, distribute, display, and perform your work – and derivative works based upon it – but for non-commercial purposes only.
ND	No Derivative Works	You let others copy, distribute, display, and perform only verbatim copies of your work, not derivative works based upon it.

The different possible combinations of the attributes lead to a licensing suite of the following six licences:

- Attribution Non-Commercial No Derivatives (BY-NC-ND);
- Attribution Non-Commercial Share Alike (BY-NC-SA);
- Attribution Non-Commercial (BY-NC);

[23] For an extensive overview of these clauses, see M. Dulong de Rosnay (2010). *Creative Commons Licenses Legal Pitfalls: Incompatibilities and Solutions*, http://www.creativecommons.nl/downloads/101220cc_incompatibilityfinal.pdf (accessed on 26/08/2011).

- Attribution No Derivatives (BY-ND);
- Attribution Share Alike (BY-SA);
- Attribution (BY).

In addition, the Creative Commons Zero (CC0) licence has been introduced, allowing one to waive all copyrights and related or neighbouring rights in one's work, such as moral rights (to the extent that these can be waived), publicity or privacy rights, rights protecting against unfair competition, database rights and rights protecting the extraction, dissemination and re-use of data. Next, the Public Domain Mark enables works that are no longer subject to copyright to be marked as such in a standard and simple way, making them easily discoverable and available to others.

Benefits and drawbacks

The main benefits of using CC licences for disseminating geographic data are that the licences have been meticulously developed and debated by leading legal scholars, they are well known and widely used across the globe, and their legal provisions have been translated into numerous languages. Major web search engines automatically pick up embedded html code indicating that the returned sites contain CC licensed material[24]. In addition, the validity of the licence has been upheld in various law suits around the world[25].

To some, one drawback of using CC licences is that they may not be altered in any way[26] (even though others see this as a great benefit)[27]. Another disadvantage of CC licences is that they cannot be used for some types of works in some jurisdictions. CC licences are intended for 'creative works' that meet the applicable legal standard for 'originality'. As this standard for originality differs between jurisdictions (e.g. in France, an imprint of the author's personality is required, while in the United Kingdom a work is original if it is the result of sufficient skill, judgment or labour[28]), geographic datasets may for instance be protected in one country, but not in another. Hence, using CC licences in a cross-border context may become complicated. The different national versions that have been created of the CC licences, all applying terminology adapted to their national legal frameworks, may complicate this even further[29].

---

[24] H. Onsrud, J. Campbell, and B. van Loenen (2010). "Interoperable Open Access Licenses for the Global Earth Observation System of Systems (GEOSS)". *International Journal of Spatial Data Infrastructures Research* 5, 194-215.

[25] See for example District Court Amsterdam, *Curry v Audax Publishing*, 9 March 2006 (the Netherlands); Court of First Instance of Badajoz, *Sociedad General de Autores e Editores*, 17 February 2006 (Spain); Tribunal of Nivelles, *Lichôdmapwa*, 26 October 2010 (Belgium). See also http://www.edri.org/edrigram/number4.6/jurisprudence-creative-commons-europe (accessed on 26/08/2011).

[26] See J. Farchy (2009). "Are Free Licenses Suitable For Cultural Works?". *European Intellectual Property Review* 31(5), 255-263.

[27] H. Onsrud, J. Campbell, and B. van Loenen (2010). "Interoperable Open Access Licenses for the Global Earth Observation System of Systems (GEOSS)". *International Journal of Spatial Data Infrastructures Research* 5, 194-215; M. Dulong de Rosnay (2010). *Creative Commons Licenses Legal Pitfalls: Incompatibilities and Solutions*, http://www. creativecommons.nl/downloads/101220cc_incompatibilityfinal.pdf (accessed on 26/08/2011); S. Dusollier (2006). "The Master's Tools v. The Master's House: Creative Commons v. Copyright". *Columbia Journal of Law & Arts* 29, 271-293.

[28] K. Janssen and J. Dumortier (2007). "The protection of maps and spatial databases in Europe and the United States by Copyright and the Sui Generis Right". *John Marshall Journal of Computer and Information Law* 24, 195- 225.

[29] M. Dulong de Rosnay (2010). *Creative Commons Licenses Legal Pitfalls: Incompatibilities and Solutions*, http://www.creativecommons.nl/downloads/101220cc_incompatibilityfinal.pdf (accessed on 26/08/2011).

In addition, problems of interpretation may occur for some of the CC licences, e.g. for the licence that only allows non-commercial use. What exactly constitutes 'commercial use'? What about a company representative visiting a client using a car navigation system: does this constitute commercial use? The vagueness of the term 'non-commercial' has been criticised by several authors[30], and while on a national level some consensus may be reached on the exact scope of the term 'commercial', on a cross-border or international level this will be much more difficult.

Moreover, the use of only a non-commercial CC licence may be a problem for geographic data stemming from the public sector in particular jurisdictions, e.g. in the European Union, where both non-commercial and commercial use of such geographic data should be allowed under the Directive on the re-use of public sector information[31]. A separate licence would still be possible for commercial use, but this would limit the harmonising potential of the CC licence.

Next, the CC licence concepts of 'no derivatives' and 'share alike' may also pose a problem if the aim is to make datasets available for added-value products. If information (including geographic data) is only to be used without the possibility to produce derivatives, then it will only be suitable for internal business processes or for end users, and the addition of value by other users is not possible. Moreover, in a creative environment the concept of sharing works, adapting them, and making the derivatives available under similar conditions, as required by the share alike condition, might be important to control potential free-riders who want to redistribute the work and its derivative works under a more strict licence[32]. However, when geographic data are made available for the purpose of value adding, the requirement of making the added-value services and products available under the same open conditions would be counterproductive to the business model of many value-adders.

Licensing frameworks in the geo-domain

Even though the calls for standardisation of licensing conditions are increasing, many public bodies providing geographic data are still reticent to move away from their existing licensing policy and their national legislation[33]. Concerns remain about the validity of international

[30] M. Dulong de Rosnay (2010). *Creative Commons Licenses Legal Pitfalls: Incompatibilities and Solutions*, http://www.creativecommons.nl/downloads/101220cc_incompatibilityfinal.pdf (accessed on 26/08/2011); F. Welle Donker, B. van Loenen and J. Zevenbergen (2010). "Geo Shared licences: a base for better access to public sector geoinformation for value-added resellers in Europe". *Environment and Planning B: Planning and Design* 37, 326-343; S. Dusollier (2006). "The Master's Tools v. The Master's House: Creative Commons v. Copyright". *Columbia Journal of Law & Arts* 29, 271-293; V. Rutledge (2008). *Fair Comment: Towards a Better Understanding of NC Licenses*, http://www.col.org/news/Connections/2008feb/Pages/fairComment.aspx (accessed on 26/08/2011).

[31] M. van Eechoud and B. van der Wal (2007). *Creative commons licensing for public sector information. Opportunities and pitfalls*, http://www.ivir.nl/publications/eechoud/CC_PublicSectorInformation_report_v3.pdf (accessed on 17/04/2011); K. Janssen (2010). *The Availability of Spatial and Environmental Data in the European Union. At the Crossroads between Public and Economic Interests*. Alphen a/d Rijn: Kluwer Law International.

[32] J. Lerner and J. Tirole (2005). "The scope of open source licensing". *Journal of Law, Economics and Organisation* 21(1), 20-56. See also K. Stewart et al. (2006). "Impacts of Licence Choice and Organizational Sponsorship on User Interest and Development Activity in Open Source Software Projects". *Information Systems Research* 17(2), 126-144, noting that restrictive licences, including viral open source licences, have a negative effect on user interest.

[33] K. Janssen, B. van Loenen and G. Remetey-Fulopp (2011). "Harmonising Licences on a Global Level: Mission Impossible or Piece of Cake?". *Proceedings of the INSPIRE conference, Edinburgh*, http://inspire.jrc.ec.europa.eu/events/conferences/inspire_2011/abstracts/240.doc (accessed on 26/08/2011).

licensing models, and the applicability of these licences to particular national or sectorial situations.

However, recently several initiatives have been taken aimed at harmonising licences for public sector (geographic) data. In this section we describe three of these initiatives: Geo Shared in the Netherlands, the Government Information Licencing Framework (GILF) in Queensland (Australia), and the INSPIRE basic and specific licence in the European Union. Many more initiatives are currently under development. For an overview of some of these we refer to the INSPIRE Good Practice Guide[34].

Geo Shared (Netherlands)

The Dutch Geo Shared licensing framework is embedded in the more general policy of the Dutch government to make public sector information available free of charge and without any (re-) use conditions (see Table 3)[35]. This policy applies to all information held by national government organisations and is endorsed by the Dutch provinces and water authorities. Most public sector information is envisioned to be available under a Creative Commons Public Domain Mark[36]. If a PDM is not possible, a CC0 declaration is advised.

Table 3: Recommended policy for Dutch public sector information

PUBLIC DOMAIN	PDM: Work is in the public domain
PUBLIC DOMAIN	CC0: Work is donated to the public domain

However, in some instances public bodies are still required to apply use conditions and/or charges, for example due to legal obligations, or because of costs that need to be recovered. For the situations where such use conditions are required for the dissemination of geographic data, the Geo Shared (Dutch: Geogedeeld) licensing framework was developed as part of the Dutch INSPIRE programme, in close cooperation with the geographic data sector in the Netherlands. The framework builds on the CC concept, including symbols representing the various use conditions, a layman's text and a legal text. The data providers can choose the use conditions they want to apply to the dissemination of their geographic data from a limited list, shown in Table 4. The symbols representing the conditions of use are published in the Dutch national geo-register[37]. In addition to the licensing framework,

[34] INSPIRE Drafting Team on Data and Service Sharing (2010). *Good practice in data and service sharing*, http://inspire.jrc.ec.europa.eu/documents/Data_and_Service_Sharing/GoodPractice_%20DataService%20Sharing_v1.1.pdf (accessed on 26/08/2011).

[35] See van R. Van Boxtel (2000). *Naar optimale beschikbaarheid van overheidsinformatie*. Minister voor Grote Steden- en Integratiebeleid. Tweede Kamer, vergaderjaar 1999-2000, 26 387, nr. 7, https://zoek.officielebekendmakingen.nl/kst-26387-7.pdf (accessed on 18/10/2011) [in Dutch].

[36] See http://creativecommons.org/licenses/publicdomain/deed.en_US (accessed on 26/08/2011).

[37] See http://www.nationaalgeoregister.nl/geonetwork/srv/en/main.home (accessed on 26/08/2011).

a licence generator[38] was developed. On 21 September 2010, the framework was accepted by the GI Council, the Dutch advisory council on geographic data. In the course of 2011, it is being implemented by the public sector data providers.

Table 4 presents the standard use conditions that can be selected in the Geo Shared licensing framework.

Table 4: The Geo Shared licensing framework[39]

	Attribution: Work can be used and re-used if name of copyright holder and/or the date of the creation of the Work is mentioned.
	Derivative works only if: the Work can only be part of a Derivative Work if the Work is not selectable from the Derivative Work.
	No redistribution: No redistribution of the Work is allowed.
	Time limitation: The licence is valid for a limited period.
	Fee required: Use of the Work requires a monetary payment.
	Purpose limitation: The Work can only be used for the purpose(s) specified in the licence.
	Additional conditions: Other restrictions than the above apply.

38 The licence generator provides an overview of the possible use restrictions. After one has chosen the restrictions that apply, the licence generator automatically processes a standard licence based on the selected use restrictions (see http://geogedeeld.geonovum.nl [accessed on 26/08/2011]).

39 See B. Van Loenen and D.W. van Barneveld (2010). "Implementing INSPIRE: the process towards the harmonization of licenses for public sector geographic information in the Netherlands". *Proceedings of the INSPIRE conference, Krakow,* http://inspire.jrc.ec.europa.eu/events/conferences/inspire_2010/abstracts/118.doc (accessed on 26/08/2011); B. van Loenen and F. Welle Donker (2010). "Towards a framework of standard licenses for geo-information?!". Presentation at the GSDI-12 Conference. Realising Spatially Enabled Societies. Singapore, www.gsdi.org/gsdiconf/gsdi12/slides/1.3a.pdf (accessed on 18/10/2011).

Government Information Licensing Framework (Queensland, Australia)

The Government Information Licensing Framework (GILF) of Queensland is the result of a project initiated by the Queensland Spatial Information Council (Australia)[40]. Its objective is to enable public sector information users to understand the rights of use associated with the material they want to use. GILF comprises a licensing framework containing the six CC licences and one GILF Restrictive Licence[41]. Originally only used in Queensland, it has now been taken up by the other Australian States and territories under the name of AUSGOAL, Australian Governments Open Access and Licensing Framework[42].

Under GILF/AUSGOAL, the six CC licences are the preferred method for licensing government IPR. However, the restrictive licence template can be used if the public bodies want to impose additional conditions. It has been developed specifically for material that contains personal or other confidential data, but it may also be used for other reasons, including for material that is licensed with limiting or restrictive conditions.

Like the CC licences, the GILF/AUSGOAL restrictive licence includes an icon, layman text and a full licence. The licence contains 30 articles with 16 additional options to specify the generic licence provisions[43]. In principle, the data obtained under this licence can be used in Australia for the own internal purposes of the user. Restrictions include prohibitions to copy the data, to make them available, to transmit them electronically or to perform any act that is not explicitly allowed under the licence. The licence provides an appendix in which these standard restrictions can be overturned, and the making of copies, the online distribution, the electronic transmission, the distribution of hard copies or anything else can be allowed. Further, the GILF licence has an article on payment and in an appendix information on the licence fee and payment terms can be provided. As already indicated, the restrictive licence also pays explicit attention to the situation where confidential and/or personal data is made available to the user. It allows for specific use conditions that can take into account the specific needs for protection of these types of data.

INSPIRE Drafting Team Basic and Specific licences (EU)

In 2005, the INSPIRE Drafting Team on Data and Service Sharing was set up to prepare the Commission Regulation implementing Article 17.8 of the INSPIRE Directive on access under harmonised conditions by the bodies and institutions of the European Community (now European Union) to spatial data sets and services from the Member States.

The Drafting Team provided a guidance document for the Member States and public authorities on how to share spatial data sets and services with the Community institutions and bodies[44]. In this guidance document, the Member States are encouraged to make

[40] See www.qsic.qld.gov.au/initiatives/gilf.html (accessed on 26/08/2011). See also the contribution of Anne Fitzgerald further on in this book on the Australian experiences in opening up access to geographic data.

[41] A. Fitzgerald (2010). "Legal Issues and Experiences from an Australia Perspective". Presentation at GSDI-12 Conference 'Realising Spatially Enabled Societies'. Singapore, 19-22 October.

[42] See http://www.ausgoal.gov.au/ (accessed on 26/08/2011).

[43] See http://www.ausgoal.gov.au/restrictive-licence-template (accessed on 26/08/2011).

[44] INSPIRE Drafting Team Data and Service Sharing (2010). *Guidance on the Regulation on access to spatial data sets and services of the Member States by Community institutions and bodies under harmonised conditions*, http://inspire.jrc.ec.europa.eu/documents/Data_and_Service_Sharing/DSSDraftGuidancedocument_v4.1.pdf (accessed on 26/08/2011).

upstream framework INSPIRE agreements on data sharing between multiple organisations and for multiple datasets, in this way preventing the need for a separate licence for each request for data. However, if such agreements are not available, the Member States are encouraged to use a Basic or Specific INSPIRE licence[45]. While these licences were created specifically for the dissemination of spatial data and services towards the EU institutions and bodies, their broader relevance for any exchange of spatial data was also kept in mind.

The Basic INSPIRE Licence applies when spatial data sets or services can be used under INSPIRE conditions without significant further restrictions, and the use is free of charge. These INSPIRE conditions hold that the data or service can be used for the performance of public tasks that may have an impact on the environment by the institutions and bodies of the Community, and by contractors on their behalf. The institutions and bodies can allow public access to the data or service, but they should avoid unnecessary duplication of the original data set or service from the data provider, or any data or service derived from it[46]. The licence also contains standard provisions on warranties and security measures, liability, access and delivery methods, personal data, assignment and sub-licensing, conflict resolution and termination.

Towards the development of cross-border and global standard geo-licences

The initiatives mentioned above show that we are starting to move from ad hoc licences created by individual organisations towards nationally or sectorially coordinated harmonised licences (see figure 2). The motivation behind these evolving developments is to improve the legal interoperability of geographic data and geographic information services. However, while these initiatives should be applauded, they still do not provide a solution for users intending to combine data from different sectors and across borders. The GSDI Legal and Socio-Economic Working Group aims to remedy this by proposing a global licensing model for geographic data.

The process of comparing and categorising licence terms

In order to create such a global model, the Working Group started with the collection of existing material on (national and international) licensing frameworks. The key components thereof were compared and categorized in a number of 'common denominator' groups. Based on these common denominators, a framework of several types of licences will be developed that can be used globally to increase the transparency of the conditions for obtaining and using geographic data. Throughout the process, the intention is to avoid creating unnecessary new licensing conditions, but also to accommodate possible differences between organisations, cultures, and financing models. An overview of the material that was compared is given in Table 5.

[45] Ibid.
[46] Ibid.

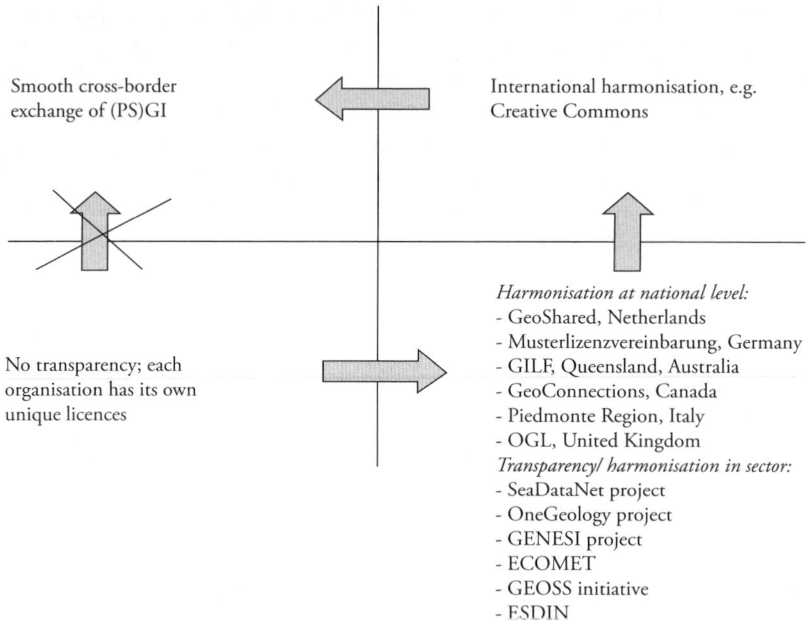

Figure 2: Suggested stages of development of a standard for geo-licences

In the review of the different licences and licensing frameworks, it was found that almost all of them contained provisions for the same topics. At first sight, this might mean that these licences could be harmonised easily, as they all discuss the same matters and the users come across the same elements in each licence. However, even though the topics might be the same, this does not mean that the content of the provisions is comparable. Hence, although agreement on general harmonisation of these categories would already be a step forward, this would only provide a small benefit to the user. Further harmonisation is needed at two levels. First, standard formulations should be available for each provision in the licence agreements. This would entail that while the organisations applying the licensing framework would still be able to determine their own requirements and conditions for the use of their data, these requirements and conditions would need to be selected from a list of standard requirements and conditions, formulated in standard clauses with harmonised wording. On a second level, not only the formulation of the conditions and requirements would be harmonised, but the types of conditions that can be imposed would be limited, in this way encouraging the greater exchange of geographic data. This second level depends on many different legal, cultural and institutional factors and appears to be very difficult to achieve. Therefore, the objective of the Working Group is set somewhere in between: a licensing framework that provides a number of model licences built from standard clauses, but that still allows data providers to choose the conditions they want to apply.

Table 5: Overview of existing national or sectoral geo-licence harmonisation efforts provided to the GSDI Legal and Socio-Economic Working Group

Name/ organisation	**Location/Region**	**URL**
APIE	France	https://www.apiefrance.fr/sections/actualites/des-conditions-generales-pour-la-reutilisation-des-informations-publiques/view
CC	World	http://creativecommons.org
ECOMET	Europe	http://www.ecomet.eu/index.php?option=com_content&view=article&id=2&Itemid=3
ESDIN	European Union	http://www.esdin.eu/project/summary-esdin-project-public-deliverables#Licence
GeoConnections	Canada	http://www.geoconnections.org/publications/Best_practices_guide/Guide_to_Best_Practices_Summer_2008_Final_EN.pdf
Geo Shared	Netherlands	http://www.geonovum.nl/diensten/gebruiksvoorwaarden
GILF	Queensland, Australia	http://www.gilf.gov.au/
INSPIRE basic & specific licence	European Union	http://inspire.jrc.ec.europa.eu/documents/Data_and_Service_Sharing/DSSDraftGuidancedocument_v4.1.pdf
Ministry of Justice	France	http://www.rip.justice.fr/1932-simplified-licence-%C2%AB-conditions-of-the-reuse-of-public-information-that-is-freely-reusable
OneGeology-Europe	European Union	http://www.law.kuleuven.be/icri/deliverables/2071G-E_WP7_D7.pdf?where=
Open database licence	World	http://www.opendatacommons.org/licenses/odbl/
Open licence	Montevideo (Uruguay)	http://monolitos.montevideo.gub.uy/resoluci.nsf/de053405568724cf832575ae004f0467/7adaf8ec8d70033b832576d60041760f
Ordnance Survey (Open Government Licence)	United Kingdom	http://www.ordnancesurvey.co.uk/oswebsite/business/licences/agreements.html
SeaDataNet	Europe	http://www.seadatanet.org/content/download/3899/29604/version/2/file/SeaDataNet+Data+Policy+.pdf

Overview of the categories of licence terms

Table 6 provides an overview of the categories of terms used in the licensing frameworks that were analysed. The categories found in almost all of the licences and licensing frameworks were: definitions, grant of licence, obligations, allowed use, use restrictions, term and termination, disclaimers (limitation on liability/warranties/indemnification), dispute resolution, governing law, jurisdiction, and form and effect of the agreement.

Review of some particular categories

For many of the categories of terms identified in the review, the objective and content are the same. However, sometimes different licensing frameworks use the same categories, but with

Table 6: Matching categories of licensing terms with existing licensing frameworks

Licensing framework/ Licence component	GILF (AU)	Geo-Shared (NL)	INSPIRE Basic (EU)	INSPIRE Specific (EU)	Geo Gratis (CA)	Open Government Licence (UK)	RIP (FR)	ECOMET	Geoconnections (CA)	APIE (FR)	Seadatanet
Definitions	X	X	X	X	X	X		X	X		X
Grant of Licence	X	X	X	X	X	X	X	X	X	X	X
Restrictions on Use/ Allowed use	X	X	X	X	X	X	X	X	X	X	X
Term and Termination	X	X	X	X	X	X	X	X	X	X	
Disclaimers	X	X	X	X	X	X	X	X	X	X	X
Dispute resolution	X	X	X	X	X	X	X	X	X		
Governing Law	X	X	X	X	X	X	X	X	X	X	
Choice of Jurisdiction	X	X	X	X	X	X	X	X	X		
Form and Effect of Agreement	X	X	X	X	X	X	X	X	X		
Payment	X	X	X	X	N/A	N/A	N/A	X	X	N/A	X
Security			X	X							
Access and Delivery			X	X				X			
Personal data processing	X		X	X		X				X	
Assignment and sub-contracting	X	X	X	X			X	X	X		
Confidential data processing	X							X			
Use by contractors	X		X	X		X		X	X		

different content (different wording, different options/subcategories within the category of terms). While ideally these options or subcategories would be limited to a minimum, a big step in the harmonisation process would already be to ensure transparency about these different terms and conditions, and to create some uniformity in their formulation. In this way the user knows what to expect when he or she wants to use geographic data.

In this section, we take a closer look at four of the categories for which several options or subcategories were defined in the model licences: grant of licence, allowed use, obligations of the user, and use restrictions.

Grant of licence

While all licences that were examined provide the user with a non-exclusive grant of rights to use the geographic data, some differences were found relating to either the territory for which the licence was valid, or the necessity of payment for the use. For instance, some of the licences of Queensland's GILF only allow the geographic data to be used within Australia. Some licences explicitly request payment, such as the French APIE's (Agence du patrimoine immatériel de l'État) licences, while others are free of charge (e.g. INSPIRE basic licence, CC). Many of the licensing frameworks that include multiple licences include both free of charge and charging licence templates.

Harmonisation of these licensing frameworks would in first instance have to include options for limitation of the grant to a specific area and the inclusion of charges. However, if the licensing framework would be applied worldwide, the area limitation would of course no longer be possible. In order to increase the use of geographic data, the standard licence would ideally also be free of charge, but this would require a considerable change in the culture and business model of many public sector providers of geographic data.

Allowed use

One of the central elements of any licence is the use the licensee is allowed to make of the data that he or she has obtained under the licence. Therefore, it is not surprising that the model licences and the licensing frameworks studied in this project contain many different variations of allowed use, adapted to the specific needs of each framework and/or data provider. These variations also include many different wordings, but can in general be reduced to a number of different types of use, mainly involving:

- Accessing;
- Viewing;
- Downloading;
- Copying;
- Distributing;
- Making derivative works.

Once again, the harmonised licensing framework will need to allow the data providers to choose from these different options, as they may need to differentiate between the possible types of use for determining their conditions, procedures and, most of all, their pricing models. However, the GSDI harmonised licensing framework can provide a considerable contribution to transparency and user friendliness of the dissemination of geographic data by harmonising the many different formulations into a limited number of standard categories.

Use restrictions

Next to the allowed use, an important part of each licence for geographic data are the acts that the licensee is *not* allowed to perform with the geographic data he or she has obtained. In the review of the model licences and licensing frameworks, a wide variety of use restrictions was found. The licensing frameworks consisting of multiple model licences generally included different combinations of these restrictions in each model, offering from very open to very restrictive licences for the data providers to choose from. Possible restrictions can be divided into a number of types:

- No sublicensing;
- No direct marketing;
- Viral clause: share-alike obligation;
- No distribution or disclosure to third parties;
- Only internal use for legal persons or private use for natural persons;
- No derivative works, only non-copy derivative works, no changes or adaptations to the original information;
- Limitation on number of copies, number of views, number of users/computers;
- Only use for a particular activity, or for a particular purpose;
- Only use by a particular group of users.

The different types of use restrictions can have a varying impact on the possibilities of the licensees to use the geographic data for the purposes for which they need it. For instance, while a prohibition to sublicense only limits the user from acting as a licensor of the data, the data can still be made public. Next, not being allowed to redistribute the data does not preclude the user from creating added value products and disseminating those, while a restriction to internal use would also prevent the latter. Another example where the extent of the use restriction can make a great difference is the purpose: a licence forbidding a particular type of use (e.g. no direct marketing) has much less impact than a licence only allowing one type of use (e.g. only for education).

As is the case for allowed use, any harmonised licensing framework will have to provide multiple options to accommodate the use restrictions that are required by the data providers. However, it deserves a more in-depth examination whether the options mentioned above usually appear in a licence in a particular combination. This could limit the options needed in the harmonised licensing framework. Yet, even if this is not the case, a benefit of the harmonised licensing framework could already be that the use restrictions are formulated in standard clauses, so that the users can easily determine what they can do with the geographic data, without having to interpret each restriction separately.

Obligations of the user

The fourth category of licence terms for which a number of different provisions and options were found is the obligations of the user. In the reviewed licensing frameworks the following obligations were found:

- No misuse of the data or misrepresentation of the data provider;
- No use of any identifiers/ trademarks of the supplier;
- Attribution;
- Notification of any misuse of the data or any infringements of the licence that were noticed by the user;
- Notifications of errors found in the data.

Again, as these subcategories were not found in all frameworks, they should be included as options in a harmonised framework, and provided under a standard wording, in order to offer more legal certainty to the users.

Discussion

Harmonising the subcategories of licence terms discussed in the previous sections might seem straightforward. However, respecting the existing licences (thus supporting all or at least most currently existing provisions) implies a considerable need for flexibility in the harmonised licensing system. For most restrictions a standard licence may be introduced, for instance an end user licence that only allows private use and prohibits (third party) disclosure of the work. For two use restrictions, 'only use for a particular activity, or for a particular purpose' and 'only use by a particular group of users' a standard option may be problematic. Distinguishing between predefined activities or use types has similar problems as the ones discussed with regard to the CC licences[47]: what is commercial use, non-commercial use, personal use, end-use? How can one delineate a sharp boundary between these different types? A similar discussion arises for user group categories: e.g. what is government use, private sector use, or non-profit use? This question is amplified when a user wants to combine different datasets. Hence, interoperable electronic licensing remains problematic in these situations. Even in a 'paper environment', challenges remain: how can a data provider make transparent what a user can or cannot do, if the criteria on which the decision is based are not defined clearly? An example of licensing models that try to tackle this is the Seadatanet project[48]. Seadatanet assigns four different predefined roles to its users. Based on these roles a user profile dictates the conditions of access to the datasets. Depending on the type of user, for some datasets access is free without any restrictions, for other datasets it is not. The applicability of this approach to other licensing frameworks needs further investigation to determine whether it meets the full interoperability requirements.

Conclusions

We believe that a global framework of standard geo-licences is a prerequisite for stimulating cross-jurisdictional use of geographic data and for successfully moving towards a service-oriented SDI in which multiple services can be integrated into new services without delay. The first step to reach this objective is to move across national and sector borders towards an international standard for licences for geographic data.

The review of existing licensing frameworks by the GSDI Working Group shows that they have many elements in common, at least at the generic level of categories. Even the more detailed subcategories show possibilities for harmonisation in most instances. The most troublesome from a harmonisation perspective, are those licences with conditions per user type, per activity and/or specifying the purposes for which a data set can be used. For these licences, transparency of terms and full, automatic interoperability will be a real

[47] F. Welle Donker, B. van Loenen and J. Zevenbergen (2010). "Geo Shared licences: a base for better access to public sector geoinformation for value-added resellers in Europe". *Environment and Planning B: Planning and Design* 37, 326-343.

[48] See www.seadatanet.org (accessed on 20/10/2011).

challenge. For all the other licences, we have only one message: start working on embedding harmonised use conditions in your licences.

We believe that the commonalities in the studied licences are such that agreement on an international framework of geographic data licences is a feasible goal that can be reached in the short term. In this chapter, we identified the commonalities between these harmonising national and sector efforts and categorised their terms. In the next phase of the work, based on these categories, a framework will be developed of several model licences that could be used worldwide and increase transparency of the conditions for obtaining and using geographic data.

Cultivating Open Geodata Ecologies: Lessons from the Implementation of the INSPIRE Directive

Prodromos Tsiavos[1], Kalliope Pediaditi[2], Konstantinos Nedas[3] and Spiros Athanasiou[4]

Introduction

On 22 September 2010, the Greek Parliament passed Law 3882/2010 entitled 'National Infrastructure for Geographic Information'[5], which constitutes the Greek national transposition law for the INSPIRE Directive[6]. This law was the result of a lengthy process that involved meticulous consultation with all stakeholders involved and an intensive effort to create a functional and sustainable system for the sharing of geographic data in the context of the Greek legal system[7].

Law 3882/2010 is an exceptionally important piece of legislation in the context of both the Greek Public Sector Information (PSI) legislative framework[8] and open data[9] in general. This is not just because it constitutes the much anticipated transposition of the INSPIRE Directive into the Greek legal system, but most importantly because it adopts a data life-cycle approach that could operate as a model for a general organisational system for the regulation of open data in the public sector. Law 3882/2010 follows data throughout their life-cycle from data production or procurement to data consumption and re-use, both by public authorities and the private sector, ensuring that sharing and re-use are possible at each stage. It is lengthy and technical legislation, comprising 35 articles, that goes beyond the scope of the Directive, appreciating the individualities of the Greek public administration and ensuring that the organisational conditions for meaningful and sustainable data sharing are established.

This chapter explores the way in which the INSPIRE Directive has been transposed into the Greek legal system by virtue of Law 3882/2010. In section 2 it examines the relationship

---

[1] London School of Economics, p.tsiavos@lse.ac.uk.

[2] Hellenic Mapping and Cadastre Organisation, kalliapediaditi@gmail.com.

[3] Ministry of Environment, Energy and Climate Change, knedas@gmail.com.

[4] Institute for the Management of Information Systems, spathan@imis.athena-innovation.gr.

[5] The Greek Law 3882/2010 uses the term geospatial information, but for consistency reasons 'geographic' is used.

[6] European Parliament and Council (2007). Directive 2007/2/EC of 14 March 2007 establishing an Infrastructure for Spatial Information in the European Community (INSPIRE), *OJ L* 108, 1.

[7] Law 3882/2010 has been the result of intensive work carried out by the Ministry of Environment, Energy and Climate Change (MEECC) between January and May 2010, which was coordinated by the Hellenic Mapping and Cadastral Organization (HEMCO) and supported by the Prime Minister's Office in collaboration and with the input from all Ministries.

[8] Law 3448/2006 implemented Directive 2003/98/EC on the re-use of public sector information in the Greek Legal System.

[9] Open Definition provides criteria for openness in relation to data, content, and software services. The Open Knowledge Definition (OKD) sets out principles to define 'openness' in knowledge – any kind of content or data 'from sonnets to statistics, genes to geodata'. The definition can be summed up in the statement that '[a] piece of knowledge is open if you are free to use, reuse, and redistribute it - subject only, at most, to the requirement to attribute and share-alike'. See http://www.opendefinition.org/ (accessed on 24/10/2011).

between Law 3882/2010 and Law 3448/2006, which implements the PSI Directive[10]. The close links between these two pieces of legislation indicate that they form part of a greater ecology regulating the use of PSI. Section 3 highlights the ways in which Law 3882/2010 goes beyond the provisions of the INSPIRE Directive and introduces a series of additional provisions that create a system for the uniform procurement of data in the public sector, their sharing between public authorities and dissemination to the general public with as few restrictions as possible. The next section describes the administrative structure which supports the objectives of Law 3882/2010. It focuses on the creation of a hybrid system of classic administrative divisions and ad hoc working groups to support the operation of Law 3882/2010. Section 5 revisits both the key provisions of Law 3882/2010 and its administrative structure to explain the overall approach as one trying to achieve openness through monitoring the entire data life-cycle. The key assumption behind this approach is that the opening up of data through licensing is the last step in a series of actions that ensure that all third party rights have been cleared and that there are no further limitations stemming from other areas of law. The final section explores the degree to which the model implemented regarding geographic data may be expanded for the creation of an infrastructure supporting the sharing and re-use of PSI in general.

Law 3882/2010 as part of the Greek PSI legislative ecology

The PSI Directive, introduced in late 2003, aims to create a European Community-wide market for digital content products and services by allowing access to PSI as raw material for their development[11]. The ultimate objective is to achieve economic growth and job creation[12], and the means of achieving this is the harmonization of the rules regarding the re-use of public documents[13]. However, the PSI Directive does not create an obligation to allow the re-use of documents[14]: it is left to the Member States or the specific public sector body[15] to decide whether re-use will or will not be authorized.

The PSI Directive's definition of a document is very broad and subject to developments in the information society: It covers any representation of acts, facts or information – and any compilation of such acts, facts or information – whatever its medium (written on paper, or stored in electronic form or as a sound, visual or audiovisual recording), held by public

---

[10] Directive 2003/98/EC of the European Parliament and of the Council of 17 November 2003 on the re-use of public sector information, *OJ L* 345, 90.

[11] Recital 1-5 PSI Directive.

[12] Recital 5 PSI Directive.

[13] Recital 6 PSI Directive.

[14] Recital 9 PSI Directive.

[15] Recital 10 PSI Directive: "The definitions of 'public sector body' and 'body governed by public law' are taken from the public procurement Directives (92/50/EEC, 93/36/EEC and 93/37/EEC and 98/4/EC). Public undertakings are not covered by these definitions. Articles 2.1 and 2.2 state that "1. "public sector body" means the State, regional or local authorities, bodies governed by public law and associations formed by one or several such authorities or one or several such bodies governed by public law; 2. "body governed by public law" means any body: (a) established for the specific purpose of meeting needs in the general interest, not having an industrial or commercial character; and (b) having legal personality; and (c) financed, for the most part by the State, or regional or local authorities, or other bodies governed by public law; or subject to management supervision by those bodies; or having an administrative, managerial or supervisory board, more than half of whose members are appointed by the State, regional or local authorities or by other bodies governed by public law".

sector bodies[16]. Re-use is defined as 'the use by persons or legal entities of documents held by public sector bodies, for commercial or non-commercial purposes other than the initial purpose within the public task for which the documents were produced. Exchange of documents between public sector bodies purely in pursuit of their public tasks does not constitute re-use'[17].

The scope of the PSI Directive is limited by the fact that public sector bodies may make available documents in any pre-existing format or language[18], but they are not obliged to make available documents requiring a disproportionate effort[19]. With regard to charging, the income of public sector bodies should not exceed 'the total costs of collecting, producing, reproducing and disseminating documents, together with a reasonable return on investment, having due regard to the self-financing requirements of the public sector body concerned, where applicable'[20].

The conditions for accessing the documents have to be made clear, and indices facilitating the identification of documents, where appropriate, are encouraged[21]. The re-use of documents could take place with or without a licence. Where a licence is used a number of clauses are suggested such as 'liability, the proper use of documents, guaranteeing non-alteration and the acknowledgement of source'[22]. The terms have to be fair and transparent, and the use of standard licences[23] is strongly encouraged. Third party intellectual property rights are respected and not affected by the PSI Directive[24].

The PSI Directive has been transposed into the Greek legal system by Law 3448/2006. The latter follows very closely the letter of the Directive, and in terms of regulatory structure it appoints the Ministry of Interior[25] as the body responsible for its implementation.

The provisions regarding licensing[26] stipulate that in cases of PSI re-use, the documents' content cannot be changed or distorted in any way and the re-user must refer to the source as well as the date it was last updated[27]. This limitation repeats Recital 17 of the PSI Directive, but should be interpreted broadly: it does not amount to a general prohibition on making changes to the original public documents; it should rather be read as a rule that seeks to

---

16 Article 3 PSI Directive: "'document' means: (a) any content whatever its medium (written on paper or stored in electronic form or as a sound, visual or audio- visual recording); (b) any part of such content".

17 Article 2.4 PSI Directive.

18 Article 5.1 PSI Directive: "Public sector bodies shall make their documents available in any pre-existing format or language, through electronic means where possible and appropriate. This shall not imply an obligation for public sector bodies to create or adapt documents in order to comply with the request, nor shall it imply an obligation to provide extracts from documents where this would involve disproportionate effort, going beyond a simple operation".

19 Article 5 PSI Directive.

20 Recital 14 PSI Directive.

21 Recital 14 PSI Directive.

22 Article 8.1 PSI Directive.

23 Article 8.2 PSI Directive: "In Member States where licences are used, Member States shall ensure that standard licences for the re-use of public sector documents, which can be adapted to meet particular licence applications, are available in digital format and can be processed electronically. Member States shall encourage all public sector bodies to use the standard licences".

24 Recital 22, 24, Article 1.2(b) PSI Directive.

25 According to Article 13 Law 3448/2006, the technical details for the implementation of the law are to be arranged by virtue of ministerial decrees issued by the Minister of Interior (since 2009 Minister of Interior, E-Government and Decentralised Administration). Article 14 stipulates that the codification of all legislative provisions regarding the further use of PSI will take place by virtue of a Presidential Decree. Neither the Ministerial Decision nor the Presidential Decree have been issued at the time of writing of this chapter (August 2011).

26 Article 8 Law 3448/2006.

27 Article 6.3 of Law 3448/2006 stipulates that documents may be re-used without being altered or distorted in any way and with explicit reference to their source and date when they were last updated.

differentiate the original documents clearly from any subsequent re-use. This interpretation is supported by the most recent provision of Article 6.4 of Law 3979/2011 regarding the informational obligations of public sector bodies, which stipulates that PSI made available by public sector bodies may be freely obtained, downloaded, stored, processed, disseminated and altered provided there is explicit reference to its source and, if it is altered, it is not presented as the original information.

The INSPIRE Directive, while very closely linked to the PSI Directive, is an instrument that has related but distinct goals: first and foremost, it creates interoperable national spatial data infrastructures to support the European Community's environmental policies[28] and to ensure the coordination and flow of information between national authorities or from national authorities to European Community institutions[29]. In that sense, it extends and expands the Directive on Public Access to Environmental Information[30]. The INSPIRE Directive, however, also aims to stimulate the provision of value added services by third parties in order 'inter alia [to] allow European companies to exploit its potential and contribute to economic growth and job creation'[31]. In that sense it shares some common goals with the PSI Directive, which is also confirmed by the fact that the INSPIRE Directive explicitly states that it is without prejudice to the PSI Directive[32].

The INSPIRE Directive is also differentiated from the PSI Directive in the sense that it focuses more on the sharing of information between public sector bodies[33] and the establishment of a common set of metadata[34] and services rather than the provision of data to the general public. Hence, it contains provisions that relate to interoperability[35] at both the national and trans-European levels. It also contains very detailed technical provisions as to how the metadata are to be structured and how the services should be offered. It also differs from the PSI Directive in the sense that it refers to data, metadata and services rather than just information.

The transposition of the INSPIRE Directive has to be seen as part of a broader effort to create a common information environment, and as a direct continuation of Law 3448/2006, which transposes the PSI Directive into the Greek legal system. This is explicitly mentioned in Article 2 of Law 3882/2010 and is reflected in a number of other provisions, particularly those referring to data sharing[36], re-use by third parties[37], and limitations on re-use[38].

The provisions on data sharing and re-use by third parties follow the same principles as those found in Law 3448/2006, particularly with respect to the limitations on *gratis* sharing of geographic data when public authorities use them for economic purposes that do not

[28] Recital 1 INSPIRE Directive.

[29] Recital 2-5 INSPIRE Directive.

[30] European Parliament and Council. Directive 2003/4/EC of 28 January 2003 on public access to environmental information and repealing Council Directive 90/313/EEC, *OJ L* 41, 26.

[31] Recital 5 INSPIRE Directive: "One of the principal aims of the establishment of an internal market is the creation of conditions conducive to the development of Community-wide services. Public sector information is an important primary material for digital content products and services and will become an even more important content resource with the development of wireless content services. Broad cross-border geographical coverage will also be essential in this context. Wider possibilities of re-using public sector information should inter alia allow European companies to exploit its potential and contribute to economic growth and job creation".

[32] Article 2.1. INSPIRE Directive.

[33] Article 17 INSPIRE Directive.

[34] Article 5 INSPIRE Directive.

[35] Article 7 INSPIRE Directive.

[36] Article 27 Law 3882/2010.

[37] Articles 11 and 28 Law 3882/2010.

[38] Article 10 Law 3882/2010.

relate to their public task. Article 28 of Law 3882/2010 essentially repeats the provisions of Article 8 of Law 3448/2006: where the public sector bodies operate outside their public task, they are treated as third parties and they have to be charged in the same way as non-public sector bodies. This means that they may be charged both for the cost of collection, production, reproduction and dissemination as well as return on investment costs[39]. In contrast, Article 26.3 of Law 3882/2010 allows only marginal cost charges between public authorities when they are acting within their public task[40].

The close relationship between Laws 3882/2010 and 3448/2006 has also been highlighted by participants in the stakeholders' consultation with respect to the level of the fees and the specification of different classes of licences for data sharing and re-use[41]. Law 3882/2010 allows for the provision of licences to third parties both with and without a fee, depending on the service level and quality of data provided[42]. Representatives from the Ministry of Interior have pointed out that because of the provisions in Article 2, paragraph 1 of Law 3882/2010[43], this law cannot contain any provisions that contradict the provisions of Law 3448/2006 with respect to the further use of PSI.

The Greek legislator approaches the nexus that Laws 3448/2006 and 3882/2010 form as one of setting limits to the restrictions that public authorities can impose with respect to the re-use of their data. The only limits with respect to re-use have to do with the cases where public authorities act as private entities: in all these articles, public bodies that act outside their public task are treated as third parties and are charged accordingly[44]. The difference between Law 3448/2006 and 3882/2010 is in relation to the conditions of re-use between public authorities: Law 3448/2006 is addressed to third parties intending to re-use PSI, whereas geospatial information in Law 3882/2010 has two different classes of recipients: (a) public sector bodies that may re-use information with no restrictions and only at a fee equalling the marginal cost of reproduction; (b) non-public sector bodies (third parties) that may re-use information after paying fees on the basis of the type of service offered. The maximum fee equals any reproduction costs in addition to the recuperation of investment.

The understanding of the Greek legislator that Law 3882/2010 forms an organic unit with Law 3448/2006 and hence constitutes part of the PSI legislative framework in Greece again becomes apparent in the provisions regarding procedural aspects[45]. These explicitly stipulate that the procedures for requesting geographic data from a public authority set out

---

[39] See also Recital 9, 14, 19, Article 6, 7, 10.1 PSI Directive.

[40] See also Article 17.3 INSPIRE Directive.

[41] The Public Consultation and the Consultation with Stakeholders was open between 20/08/2010 and 15/09/2010. MEECC and the Prime Minister's Office received 65 responses, 63 directly as a response to the consultation and 2 via the web site www.geodata.gov.gr which is administered by the Prime Minister's E-Government Team. In addition, a number of meetings with stakeholders took place in MEECC's premises. (S. Athanasiou (2010). *Report on the INSPIRE transposition public consultation.* Athens: MEECC).

[42] See Article 28.2 of Law 3882/2010: A service allowing the downloading of geospatial data at a minimum guaranteed time of 0.2 msecs will be charged, whereas a service without a guaranteed response time will be free of charge.

[43] This article transposes Article 2 of the INSPIRE Directive, which explicitly states that it is without prejudice to Directives 2003/4/EC and 2003/98/EC.

[44] Articles 7(1) and 11(1) Law 3448/2006 and Articles 27(8) and 28 Law 3882/2010.

[45] Cases b and c of paragraphs 2 of Article 21 Law 3882/2010.

in Law 3448/2006 have to be followed. Law 3882/2010 actually uses the same language as Law 3448/2006, which reflects the relevant provisions of the PSI Directive[46].

All these provisions contribute to an understanding of Law 3882/2010 as an integral part of the PSI legislative ecology in Greece. The Greek legislator has seen the transposition of the INSPIRE Directive as an opportunity to push further the boundaries of PSI legislation towards fewer restrictions in data sharing and re-use of data, and to establish a model for the way in which public data in general could be regulated so as to remain open throughout their life-cycle.

Elements of innovation in Law 3882/2010

The new law had to tackle some basic procedural issues, such as the setting of clear rules regarding (a) the ways in which data sets enter the public sector, particularly in relation to the clearance of rights; (b) the appointment of specific entities responsible for the curation and maintenance of the data; and (c) the setting up of a body that will provide specific rules regarding the sharing and dissemination of geographic data. Further, the national legislator consistently opted for the most open schemes for data sharing, trying to minimize fees in an effort to boost the development of value added services and to reduce transaction costs for the public administration[47].

In this context, Law 3882/2010 covers a wider range of data than the data themes covered in the Annexes to the INSPIRE Directive: (a) all datasets of the public sector that contain geographic data or have an implicit geographic reference even when the information is in a descriptive form (e.g. addresses); and (b) all spatial datasets, including those not directly relevant to the environment[48]. This is a conscious decision that again aims at the widest possible opening of geographic data in order to achieve a uniform and coherent national policy for dealing with such types of data, and to overcome the current fragmentation of practices and data sets.

The transposition of the INSPIRE Directive is frequently associated with legislation regarding intellectual property rights (Law 2121/93), personal data protection (Laws 2472/1997 and 3471/2006), PSI (Law 3448/2006) and access to public information (Article 5 of the Greek Code of Administrative Procedure). This is evident in a number of articles[49] of Law 3882/2010, which make explicit reference to such legislation, following the relevant INSPIRE and PSI provisions.

[46] See Article 4.2 of the PSI Directive: "Where no time limits or other rules regulating the timely provision of documents have been established, public sector bodies shall process the request and shall deliver the documents for re-use to the applicant or, if a licence is needed, finalise the licence offer to the applicant within a timeframe of not more than 20 working days after its receipt. This timeframe may be extended by another 20 working days for extensive or complex requests. In such cases the applicant shall be notified within three weeks after the initial request that more time is needed to process it".

[47] The innovations introduced with Law 3882/2010 are explained in the Greek INSPIRE 2010 Implementation Report and the Law 3882/2010 Supporting Documentation. These documents make explicit reference to the need to ensure that there is legislative provision for the clearance of IPR before their acquisition by public authorities, the appointment of single 'owners' responsible for the curation, acquisition and reporting regarding original geospatial datasets, and the use of registries for monitoring the acquisition, procurement, production or updating of such data sets.

[48] Article 4 Law 3882/2010.

[49] For Personal Data Protection see e.g. Articles 34 and 23.5(c), for IPR see e.g. Article 2.2 and for PSI see e.g. Article 2.1 Law 3882/2010.

In addition, Law 3882/2010 places great importance on the implementation of the interoperability framework of Law 3731/2008, which is necessary in order to ensure that a working National Data Infrastructure is in place.[50] Furthermore, Law 3882/2010 was drafted against the background of some fundamental changes occurring in the Greek public administration, particularly Law 3852/2010 (also known as project 'Kallikrates', which substantially changes the way in which local government operates, through an extensive scheme of consolidation of Prefectures). Law 3852/2010 organizes the country into 7 regional decentralized administrations, 13 prefectures and 325 municipalities. Each of these levels now has a different and broader set of powers than the previous administrative levels: for instance, the regions have more decisive and more extensive powers that include the provision of geographic data for the materialization of their public task.

Such fundamental reforms made the census of existing geographic data sets rather problematic[51], but also gave a strategic direction towards a more integrated and horizontal approach to the way in which data are managed in the Greek public sector. It is, thus, important to view the transposition of the INSPIRE Directive as part of a greater ecology of laws aiming at the cultivation of a broader common information environment.

Basic structure

The transposition of the INSPIRE Directive was conducted by the Ministry of Environment, Energy and Climate Change (MEECC)[52]. For the production of the INSPIRE implementation report, MEECC collaborated with the Open Government Team at the Prime Minister's Office[53] and the Hellenic Mapping and Cadastral Organisation (HEMCO)[54], which was assigned the role of national contact point in accordance with the INSPIRE Directive and Law 3882/2010.

Before the transposition, three studies were conducted in order to assess the optimal coordination infrastructure necessary for the implementation of a geospatial data coordination scheme: two by HEMCO, in 2008[55] and 2009[56] respectively, and one by the Technical Chamber of Greece in 2009[57]. A key concern that appeared in the studies relates to the fragmentation of geographic data ownership in the Greek public sector: while INSPIRE advocates a horizontal coordination structure, it is necessary to take into consideration the current multi-level hierarchical situation in the Greek public sector. Hence, the legislative solution adopted in Greece would have to establish a well-functioning mechanism of

[50] The existence of a coherent and operational national interoperability framework is directly linked to both Law 3448/2006 and Law 3882/2010, as it is essential in order to meaningfully provide open access to content and data produced or procured by public authorities. The Ministerial Decree necessary for the implementation of the National Interoperability Framework and the related circulars by the Minister of Interior, stipulated by Article27 of Law 3731/2008, are still to be issued.

[51] Due to the lack of comprehensive information systems infrastructures in the previous administrative units, in the process of transition a great deal of information regarding the kind and number of GIS has been lost. For instance, most of the pre-existing prefectures had no centralised index of their geographic data and thus no clear idea of what they owned. This has led to the introduction of specific census provisions in the eGovernment Law in summer 2011 (Law 3979/2011).

[52] See http://www.ypeka.gr/ (accessed on 28/10/2011).

[53] See http://www.opengov.gr/ (accessed on 28/10/2011).

[54] See http://www.okxe.gr/ (accessed on 28/10/2011).

[55] HEMCO (2008). *Inventory of Map Archives of Agencies and Bodies of the Public Sector.* Athens: HEMCO.

[56] HEMCO (2009). *National Geographical Data Infrastructure Technical Report*. Athens: HEMCO.

[57] Technical Chamber of Greece (2009). *Report on Monitoring the INSPIRE Directive: Proposal for the Implementation and Management of the INSPIRE Directive in Greece.* Athens: Technical Chamber of Greece.

coordination and at the same time somehow represent the multiplicity of stakeholders with often diverse interests and objectives.

This has been achieved with the establishment of a three-level structure. First, the National Geo-information Committee (NGC) is comprised of the General Secretaries of the most pertinent Ministries regarding the implementation of the INSPIRE Directive. It is answerable to the Prime Minister and chaired by the Minister of MEECC, and it constitutes the high-level decision making body for the geographic data infrastructure. Second, HEMCO is the operational arm of the structure, responsible for the technical implementation and coordination of the infrastructure. Third, Focal Points are responsible for the procurement, monitoring and preservation of data in accordance with Law 3882/2010. Focal Points are ad hoc working groups that do not fall within the official hierarchy, but have power for specific topics such as the implementation of the INSPIRE Directive. While Focal Points are working groups, they do not have a limitation as to their duration and their members are unpaid. The following figure, used in the Member State 2010 Report on the implementation of the INSPIRE Directive, describes the coordination infrastructure of the INSPIRE implementation in Greece:

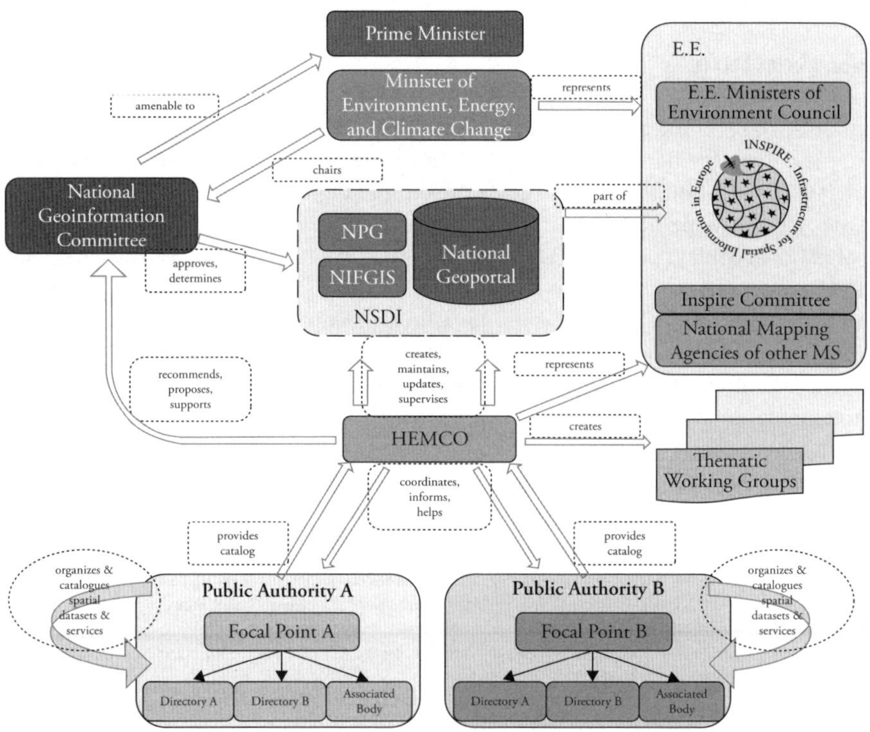

Figure 1: The administrative structure for the coordination of the National Spatial Data Infrastructure (NSDI)[58]

[58] K. Pediaditi, S. Athanasiou, K. Nedas, E. Grigoriou, and K. Stefanakis (2010). *INSPIRE Transposition. Member State Report: Greece.* Athens: MEEC.

NGC performs four key roles. First, it establishes, monitors and evaluates the national policy on Geo-information (NPG) and the National Interoperability Framework on Geographic Information and Services (NIFGIS) throughout the public sector. Next, it is responsible for taking coordination measures for the implementation of the law, approving the relevant guidelines and technical specifications. Third, it has a decision-making role regarding sharing arrangements and disputes on dataset availability. Finally, it is responsible for the review and approval of the annual monitoring and reporting provisions with regard to the INSPIRE Directive.

HEMCO is responsible for the development, operation and supervision of the National Spatial Data Infrastructure (NSDI), which comprises the NPG and NIFGIS, and of the National Geoportal. Next, it is assigned the tasks of coordination and quality assurance of geographic datasets and services developed by or on behalf of the public sector, and awareness raising and information provision campaigns with regard to the implementation of the INSPIRE Directive. Finally, HEMCO has to coordinate and write up the monitoring and reporting requirements, as well as collaborate with other Member State contact points for the sharing of best practices.

It is important to note that in order to achieve the maximum take-up of the NPG and optimal representation of different stakeholders, HEMCO creates and coordinates working groups comprising various actors, ranging from geographical information scientists and non-governmental organisations to users or producers of geographic data and governmental agencies.

Focal Points (Article 19) are set up in each public body and are responsible for the cataloguing, procurement, production, maintenance and updating of all geographic data and services of the public body. Focal Points also provide HEMCO with the census of existing datasets and services for which they are responsible.

This low level infrastructure is quintessential for the operation of the law, and is further complemented by the setting up of 'Geospatial Data Owners'[59]. These are public authorities responsible for the original procurement, production, continuous updating and preservation of an original geographic data set. An original data set is the one regarded as unique in the public administration. Where there are multiple identical datasets, one is selected as the original. Geospatial Data Owners and Original Geospatial Datasets are defined as such by NGC on HEMCO's recommendations.

Establishing Geospatial Data Owners is extremely important in the Greek public administration context as it allows a very clear allocation of responsibility for specific data sets. In previous efforts to release PSI, as with Law 3448/2006, the lack of association of specific data sets with specific organisations prevented the objectives of the law from being achieved: no organisation would assume responsibility for the maintenance or provision of a specific data set unless it was specifically stipulated in the law, which in most cases was not the case. It is important to note that an 'owner' for the purposes of Law 3882/2010 is more a custodian, an entity responsible for ensuring the data set's compliance with the INSPIRE framework, rather than an intellectual property rights holder, although the two

[59] Articles 14 and 21 Law 3882/2010.

roles may coincide[60]. A Geospatial Data Owner may retain the data in its own repository and is responsible only for providing HEMCO with a list of the data sets it has under its control.

A data life-cycle approach

Perhaps the most important innovation introduced by Law 3882/2010 is its life-cycle approach to geographic data and its appreciation of the problems the Greek public administration faces with respect to the procurement or production of data. While sharing and re-use of data remains the final objective, the social and economic value of such freedoms can be maximized only if there is no waste of resources in the procurement or producing of data. By allowing a single point of procurement for geographic data, the risk of multiple procurements of the same data set under exclusive licences by many public authorities with no central coordination has been substantially reduced. The lack of an index of geographic data in the Greek public administration was the main reason behind the introduction of this single point of procurement. In addition, for the social and economic value to be maximized, all necessary rights or licences need to have been obtained so that sharing and re-use are actually possible.

A look at the current Greek legislative framework with respect to PSI (including Law 3448/2006, the Greek Access to Administrative Documents legislation[61], the right to place information requests[62] and the right of access of the data subject to his or her personal data[63]) highlights some of the risks of not mandating rights clearance procedures for content or data procured by the public administration[64]. The Access to Administrative Documents legislation does not make any reference to re-use, and PSI legislation excludes all information on which third parties hold IPR[65]. The INSPIRE Directive also excludes content on which third parties hold IPR. While such restrictions are not necessarily problematic, they have proven to be a major obstacle to the release of geographic data, in particular in the Greek context.

This is due to the fact that most of the geographic data involved, e.g. maps, are copyrighted works under the Author's Rights Law, Law 2121/1993[66], and are thus granted copyright protection. In addition, it is standard practice that such maps are either produced by specific agencies or procured by the Greek public administration from third parties. In both cases, while there could be an easy-to-follow procedure for acquiring the rights or licences necessary further to share or re-use the data, no such provision is made. As a result, most of such content or data acquired by the Greek public sector lacks the necessary licences to be freely shared and is excluded from the scope of both the PSI legislation and the INSPIRE transposition.

---

[60] The use of the term 'owner' has raised significant concerns with respect to its meaning. Participants to the public consultation have highlighted that it may be taken to mean the owner of the IPRs. However, as the Law 3882/21010 Supporting Documentation clearly states and is explained in the definition of Article 14(2) the Owner is effectively the public authority that acts as a custodian of a data set responsible for its original procurement, production, continuous update and conservation.

[61] Article 5 Law 2690/1999 'Hellenic Code for Administrative Procedure'.

[62] Article 4 Law 2690/2010.

[63] Article 12 Law 2472/1997.

[64] S. Vlachopoulos (2007). *Transparency in the State's Actions and Personal Data Protection.* Athens: A.N.Sakoulas.

[65] Article 3 paragraph 2b Law 3448/2006.

[66] Art.2(1) Law 2121/1993.

This issue came up several times during the open public consultation, the consultation with the stakeholders, and the surveys conducted by MEECC, HEMCO and the Prime Minister's Office, and during an open call for suggesting data sets the citizens would like to see open. In all cases, the issue of not having access to data sets that have been paid for with public money was repeatedly raised, particularly with respect to geographic data.

Law 3882/2010 offers a solution to this problem by mandating a single point of responsibility for each original data set and by obliging the Geospatial Data Owner, whenever it decides to procure, produce or update a geospatial data set, to use the National Geoportal to check that the data set has not been obtained by another public authority[67]; to inform HEMCO about its decision to procure, produce or update the dataset; and to obtain all rights and licences necessary for freely sharing the data with other public agencies or allowing re-use by third parties under appropriate licensing schemes[68]. The principle is that there is a minimum level of freedoms that the public authorities and the third parties are entitled to enjoy. For instance, public authorities performing their public task acquire a licence that allows them to exercise all the economic rights stipulated by the Author's Rights Law, the only restriction being that they cannot sell these data to third parties without the permission of the data owner. Next, the general public is entitled to have access to search and view services at no charge, unless a minimum level of service is guaranteed. The data owner has to make geographic data available to the re-user but is allowed to choose the type of licence with which they are made available. Metadata are available to all with no restrictions other than attribution of the source[69]. In all cases where public money is spent, the rights or licences acquired by the Greek public administration have to be equal to or greater than the rights or licences that they grant to other public authorities or third parties. This ensures that the public bodies will not license rights that they have not acquired. The openness of the data is further established by ensuring that the provision of any geographic data-related service is in accordance with the National Geospatial Data Interoperability framework and the National Geospatial Data Policy[70].

Article 27 of Law 3882/2010 provides the framework for geographic data sharing, while the specifics of the licensing are to be defined in the National Geospatial Data Policy. It is important to highlight that Article 27 goes beyond the minimum in the INSPIRE Directive, and allows the gratis sharing between public organisations of metadata and services, irrespective of whether or not they are related to the environment. Exceptionally, it allows for charging if necessary, but only to cover the reproduction and dissemination costs, thus setting an extremely low threshold and making the geographic data available at virtually no charge for the public administration, as long they act within their public task.

Paragraph 7 of Article 27 introduces an important provision with regard to the harmonization of the licensing schemes used for data sharing. The objective of this provision is to ensure that either a single licence is used for all similar uses, or, if the administration opts for different open licences, that these are compatible with each other and we are not faced with the phenomenon of the 'fragmentation of the commons'[71] or the 'commons

[67] Article 19 Law 3882/10.
[68] Articles 20 and 21 Law 3882/10.
[69] Articles 26, 27 and 28 Law 3882/2010.
[70] Articles 22-25 Law 3882/10.
[71] N. Elkin-Koren (2006). "Exploring Creative Commons: A Skeptical View of a Worthy Pursuit". In Hugenholtz, B. and Guibault, L. (ed.). *The Future of the Public Domain*. Alphen a/d Rijn: Kluwer Law International, http://ssrn.com/paper=885466 (accessed on 24/10/2011).

archipelago'[72]. It is frequently the case that, while different public organisations choose to release their content under open licences, the lack of central coordination prevents the licences from interoperating and hence creates multiple islands of content that cannot be legally mixed, irrespective of the relevant legislation and the choice of open licences. This is particularly the case with copyleft licences that impose conditions upon the use of derivative works. Copyleft licences allow the licensee to make derivatives on the condition that these derivatives are further disseminated under the same terms and conditions as the original licensed work. As a result, if two works are licensed under different copyleft licences, a derivative work using both works may not be possible (e.g. you cannot legally make a derivative work from a Creative Commons Attribution Non-Commercial Share Alike database and Open Data Commons Attribution Share Alike database, as they do not have the same terms and conditions and their copyleft provisions conflict with each other). Thus, the best strategy is to adopt licences with the least possible restrictions, such as the Creative Commons Attribution or the Creative Commons Zero licences. Such licences are able to create the maximum leverage for the re-use of the content, while reducing transaction costs and minimizing the risk of commons fragmentation. This is possible because they do not contain share alike provisions, and hence derivative works can be created without legal frictions like those illustrated above.

Article 28 of Law 3882/2010 sets the conditions for the re-use of public geographic data by third parties, i.e. anyone not included in the definitions of public authorities. A distinction is made between metadata, data and services. Metadata are provided without charge, while the data are disseminated with licensing terms depending on their use. Finally, the download, transformation and invoke services are provided with terms that depend both on their purpose and the service levels provided. With respect to the charging of fees in the case of both data and services, the public authorities are obliged to make data available both with licences for a fee and free of charge. This construction assumes that there are criteria for making the distinction. Such criteria are implied in Article 28, but are fully defined in the National Geospatial Data Policy. The fact that the law refers to the purpose of use could imply a differentiation on the basis of the commercial character of use, since this is one of the cases where charging is a common practice[73], and it is found standardized in the most widespread standard licences, i.e. the Creative Commons Non-Commercial licences (CC NC). There are two reasons why Law 3882/2010 is reluctant to make a direct reference to commercial use: first, there seems to be a problem with the consistency of definitions of what amounts to commercial use, thus introducing unnecessary uncertainty and possibly making users reluctant to use a non-commercial licence, in this way reducing the potential for the further use of geospatial data. Second, while dual licensing with the use of non-commercial licensing schemes is a common approach when charging is involved, there may be other models, and an explicit reference to the non-commercial element may reduce innovation in future business models. Dual licensing with the NC elements could involve the provision of a CC licence with the NC element for no fee and a licence allowing commercial licences for a fee. Besides the definition risks that the explicit reference to NC would have, it

[72] P. Tsiavos and N. Korn (2009). *Case Studies for The Mapping of Content, Value and Rights Across the Public Sector*. London: Strategic Content Alliance.

[73] See for instance statistics for the use of CC licences that indicate that the CC BY NC SA is the most popular one. Some of the most widely quoted problems include: (1) lack of clarity as to what NC means within and across jurisdictions and subsequently (2) exclusion of a large number of resources, such as Wikipedia, which is licensed under CC BY SA and (3) fragmentation of the commons.

would also exclude business models based on the provision of a guaranteed service level. In such a model, the data provider offers a dual licensing scheme where the licensee is offered (a) a licence with no service level guarantee for free; and (b) a licence with a guaranteed service (e.g. response time, data quality) or allowing commercial use for a fee.

The dual licensing model is 'business model agnostic': the Greek legislator does not choose to charge on the basis of e.g. commercial or non-commercial use, but leaves the choice of the appropriate licensing scheme to the NGC following the suggestions of HEMCO and the agreement of the competent Ministers. Such an approach sets a minimum level of rights for the licensees and avoids the fragmentation of commons, while allowing the specific licence terms to be dynamically and continuously chosen and expressed in the National Geospatial Data Policy.

Finally, another innovation introduced by Law 3882/2010 that goes beyond INSPIRE relates to the procedure necessary to compile documents with regulatory effect that require the use of geographic data. According to Article 32, such procedure has to be completed using electronic documents, and in case of dispute the digital copy is deemed to be the original. This constitutes a great step forward in ensuring that the procedure is done with the minimum transaction cost, and maximizes the efficiency of the process.

Beyond INSPIRE

Law 3882/2010 goes beyond the INSPIRE Directive to suggest a broader but solid model for implementing open data regulatory frameworks in jurisdictions with fragmented IPR policies and moderate levels of public organisational capacity.

Organisationally, it creates a three-layered structure allowing the representation of multiple stakeholders (composition of the NGC, introduction of Thematic Networks by HEMCO, decentralized Focal Points and Geospatial Data Owners) and flexible decision making (regular reviews of the NPG), while ensuring a strong operational arm (HEMCO) and oversight at the highest level (Prime Minister and MEECC). This mixture allows both space for the different stakeholders and a strong steering of the open geographic policy making.

In terms of regulatory architecture, Law 3882/2010 wisely does not start from the point of data dissemination and re-use, but rather at the point of the production or procurement of the data. It ensures that there is a single point where data are collected, produced, procured or updated and that information about the collection, production and storage of the geographic data is registered at a central point (HEMCO). This is further complemented by strict rules prohibiting multiple purchases of the same data set and mandating the clearing of rights for all data sets acquired with public money. This is a structure that is not found in most Greek laws, because it is generally assumed that such structure may be implemented through organisational decisions, circulars or other administrative measures. This may be true for administrations at a mature organisational level, but is not something that is consistently found in all jurisdictions or in all public organisations within the same jurisdiction. A mechanism such as Law 3882/2010 thus ensures that any PSI re-use legislation can be meaningfully implemented and does not become practically devoid of content because of the actual volume of data that might be exempted due to an inconsistent rights acquisition policy across the public sector.

The provisions relating to the sharing and re-use of data also create a balance between a high standard for allowed use of the licensed material, standardization of the licensing schemes and the introduction of flexibility in the choice of licence. Whereas free choice is the key concern in most national implementations of PSI-type legislation[74], this is normally not counterbalanced with standardization provisions or minimum levels of rights for the re-users of information. The approach of Law 3882/2010 sets the virtually *gratis* licensing of geographic data for sharing between national public authorities as the minimum, and allows the choice of the specific licence to be defined in the national policy for geographic data. Most interestingly, it stipulates that for the re-use of the data by third parties both *gratis* and paid-for licences should be available. The differentiation between licences and the level of the fees is left open to the public authorities to decide on the basis of the cost recuperation scheme, the purpose of the use and the service level provided. Finally, the focus on creating a common information environment through the use of compatible licences is a very crucial feature of Law 3882/2010. Again this is something that in some jurisdictions may be achieved through administrative measures. However, its inclusion in the actual law ensures that maximum harmonization may be achieved.

Overall, Law 3882/2010 introduces a number of innovations that could be used in order to maximize the effect of PSI-type legislation, particularly when administrative measures are inadequate because of the legal culture or the capacity of the administration and the broader public sector to regulate itself. In this effort to cultivate a common information environment, Law 3882/2010 is not an isolated legislative intervention. It is complemented by a number of additional initiatives taken by the legislative[75] and the administrative level[76] that constitute positive efforts to increase transparency and support development through the use of open content.

In addition, the Special Secretary for Digital Planning[77] has introduced interoperability, open formats, standards and data as well as the degree of implementation of Law 3882/2010 as evaluation criteria for all projects funded by the Digital Convergence Program. The Special Secretary for Digital Planning is a consistent proponent of open data and the re-use of PSI as a tool for leveraging sustainable development in Greece. He has recently issued two Calls[78] that provide funding for the development of services that support the implementation of both the PSI and INSPIRE Directives through Laws 3448/2006 and 3882/2010. This is yet another indication of the approach followed by the Greek public authorities to the PSI and INSPIRE Directives as parts of the same public policy objective and essential tools

[74] See e.g. Office of Public Sector Information (2007). *The United Kingdom Implementation of the European Directive on the Re -use of Public Sector Information - the first two years* London. The National Archives.

[75] E.g. the introduction of the transparency project 'Diaugeia' (Law 3861/2010) that mandates the publishing of all decisions of Public Authorities of all levels.

[76] E.g. the Prime Minister's decision No. Y255 on setting Creative Commons Attribution GR 3.0 as the default licence for the Prime Minister's web site (http://www.primeminister.gr/) or the open geographic data website run by the Prime Minister's Open Government Team(www.geodata.gov.gr). Geodata.gov.gr is perhaps the most important implementation initiative of Laws 3882/2010 and 3448/2006. It aims at allowing the request, publication and search not just of open geographic data but all categories of data and it is meant to operate as a testbed for the opening of all public data.

[77] See http://www.infosoc.gr/infosoc (accessed on 24/10/2011).

[78] http://www.infosoc.gr/infosoc/el-GR/TodayNews/nees_proskliseis_30-09-2010.htm (accessed on 24/10/2011).

for supporting the creation of value added services and a vibrant market based on public information and data[79].

The success of Law 3882/2010 is yet to be tested in its implementation. Its importance, however, as a model for increasing administrative capacity in dealing with open data is hard to dispute.

[79] The Special Secretary of Digital Planning, Dr. Antonis Markopoulos, has issued a statement regarding the Calls 20.2 (Ministry of Interior) and 24 (Environment) that highlight the aforementioned points: "The two new calls of the Operational Program of Digital Convergence come to address significant issues in two key areas: On the one hand, in the sensitive area of Environment, where the creation of Forest Maps and the implementation of the European Directives will contribute to sustainable development and, on the other hand, in the multifaced and essential area of Public Administration and Electronic Government. The effort for the Public Sector reform and improvement of the service to the citizen is focused and sustained and for that reason our interventions are capable of bringing sustainable and substantial results." See http://www.infosoc.gr/NR/rdonlyres/35F07CFB-9BF2-4FEE-B6AA-FB0476E98BED/8139/2010_09_30_Programmatapsifiakisiglisi.doc (accessed on 24/10/2011).

Juggling Information Policy, Rights to Information and Copyright Licensing to Enhance the Accessibility and Reusability of Spatial Data: The Australian Experience

Anne Fitzgerald[1]

Introduction

There is still no comprehensive information strategy governing access to and re-use of public sector information, applying on a nationwide basis, across all levels of government – local, State and federal – in Australia. This is the case both for public sector materials generally and for spatial data in particular. Nevertheless, the last five years have seen some significant developments in information policy and practice, the result of which has been a considerable lessening of the barriers that previously impeded the accessibility and reusability of a great deal of spatial and other material held by public sector agencies. Much of the impetus for change has come from the spatial community, which has for many years been a proponent of the view 'that government held information, and in particular spatial information, will play an absolutely critical role in increasing the innovative capacity of this nation'[2]. However, the potential of government spatial data to contribute to innovation will remain unfulfilled without reform of policies on access and re-use and of the pervasive practices of public sector data custodians who have relied on government copyright to justify the imposition of restrictive conditions on the use of their data.

Breakthroughs were forged from 2005 on, catalysed in the context of an applied research collaboration between the Queensland Government[3] and legal academics at Queensland University of Technology (QUT)[4], which was established to investigate policy, legal and economic issues relating to flows of spatial data. An early outcome of this project was the proposal for a Government Information Licensing Framework (GILF) to provide access to public sector information (PSI) under an open content copyright-based licensing regime, applying the Creative Commons (CC) suite of licences to standardize the conditions for use and re-use of licensed materials[5]. This approach is based on the recognition that re-use of public sector materials can be facilitated by directly addressing the obstacles presented by

---

[1] Law and Justice Research Centre, Queensland University of Technology, am.fitzgerald@qut.edu.au.

[2] Australian Spatial Consortium (2008). *Submission 307 to the Review of the National Innovation System*, http://www.innovation.gov.au/Innovation/Policy/Pages/ReviewoftheNationalInnovationSystem.aspx (accessed on 18/10/2011).

[3] The Queensland Government participation was led through the Queensland Treasury's Office of Economic and Statistical Research headed by Dr Peter Crossman. From 2007 to 2010 the project was funded by the Cooperative Research Centre for Spatial Information.

[4] The QUT team was led by Professor Brian Fitzgerald and Professor Anne Fitzgerald. Brian Fitzgerald is also the project lead for Creative Commons. See http://creativecommons.org.au (accessed on 18/10/2011).

[5] Queensland Spatial Information Office (2006). *Government Information and Open Content Licensing: An Access and Use Strategy*, http://eprints.qut.edu.au/32117/1/Stage%202%20Final%20Report%20PDF%20Format.pdf (accessed on 18/10/2011).

the complex, protracted and restrictive licensing practices and procedures that are typically encountered[6]. These insights rapidly gained widespread acceptance and influenced subsequent developments in policy and practice throughout Australia. In practice, the GILF project led to the use of CC licences, not only as an operational mechanism for managing government copyright but also – to the extent that they ensure at least a degree of re-use of licensed materials – an element of information policy. By releasing their materials under non-exclusive, open content licences, government agencies have adopted a policy position that, by default, materials made available for access can also be used and re-used (whether by members of the public or other governmental entities) with no – or few – restrictions.

Following the review of the national innovation system in 2008[7] and the inquiry of the Government 2.0 Taskforce in 2009[8], the federal government adopted a position on licensing of government materials which goes beyond that envisaged by GILF. Responding to the recommendations of the Government 2.0 Taskforce[9], the federal government accepted the proposal that the CC Attribution (CC BY) licence should be the default licence applied when government-owned copyright materials are released for distribution[10]. Whereas the GILF model encompassed the entire suite of six Australian CC licences, together with a model licence and template clauses for use when restrictive licensing conditions were to be imposed, the federal government accepted that the only condition that would typically be applied is that the material is correctly attributed.

Contemporaneously, Freedom of Information (FOI) regimes introduced from the early 1980s onwards were being reformed to require governments to adopt pro-disclosure practices based on a 'right to information', so that a much more extensive range of information is routinely and proactively released. Synergistically, the FOI reforms and the adoption of CC licensing have led to the current position in which not only is a great deal more government information available for access than was previously the case, but it is also, as a default, available for use and re-use, subject only to attribution requirements. The Australian experience has been that the adoption of open content licensing practices to overcome copyright-based restrictions has enabled governments to improve the accessibility and reusability of their materials, in advance of the development of comprehensive and detailed information policy frameworks.

---

[6] For a detailed analysis of the restrictions that are typically applied to government materials see Queensland Spatial Information Office (2006). *Government Information and Open Content Licensing: An Access and Use Strategy*, http:// eprints.qut.edu.au/32117/1/ Stage%202%20Final%20Report%20PDF%20Format.pdf (accessed on 18/10/2011).

[7] The review of the national innovation system, chaired by Dr Terry Cutler, was commissioned by the Minister for Innovation, Industry, Science and Research, Senator Kim Carr on 22 January 2008 and reported on 29 August 2008. See Cutler & Company (2008). *Venturous Australia - Building Strength in Innovation. Report on the Review of the National Innovation System, prepared for the Australian Government Department of Innovation, Industry, Science and Research*, http://www.innovation.gov.au/Innovation/Policy/Pages/ReviewoftheNationalInnovationSystem.aspx (accessed on 18/10/2011).

[8] The Government 2.0 Taskforce, chaired by Dr Nicholas Gruen, was established in July 2009 and reported in December 2009. See Government 2.0 Taskforce (2009). *Engage: Getting on with Government 2.0*, http://www.finance.gov.au/publications/gov20taskforcereport/doc/Government20TaskforceReport.pdf (accessed on 18/10/2011).

[9] Government 2.0 Taskforce (2009). *Engage: Getting on with Government 2.0*, http://www.finance.gov.au/publications/gov20taskforcereport/doc/Government20TaskforceReport.pdf (accessed on 18/10/2011).

[10] Note that, unless otherwise indicated, references to CC licences refer to the Australian suite of CC licences. See http://creativecommons.org.au (accessed on 18/10/2011). For the CC Attribution 3.0 Australia (CC BY 3.0) licence, see http://creativecommons.org/licenses/by/3.0/au/deed.en (accessed on 18/10/2011).

Towards a national information strategy

In contrast with developments in the United States and Europe, until recently little attention has been given in Australia to the importance of developing information policy frameworks[11]. Early in the Internet era, the advantages to be gained from re-use of government held materials in new digital content and services was highlighted in *Commerce in Content: Building Australia's International Future in Interactive Multimedia Markets*[12], a report produced by Cutler & Company for the federal government in 1994. It made several recommendations as to how governments, at both federal and State level, could leverage 'off their position as significant users, custodians, and producers of content' to accelerate the development of the digital content sector[13]. The recommendations included providing easy access to culturally significant data in digital form, as well as providing comprehensive access to nationally significant data, and promoting the development of standards for document and image digitalisation and archiving. The extensive scale and variety of the cultural and content materials held by the public sector meant that '[i]n terms of scale, government initiatives would represent the biggest single catalyst for accelerated industry development'[14]. In the same year, the Australian Science and Technology Council[15] proposed that government should stimulate public interest in, and facilitate access to, government information via electronic networks.

Unfortunately the momentum dissipated and during the ensuing decade Australia lagged 'behind many other advanced countries in establishing institutional frameworks to maximise the flow of government generated information and content'[16]. This period saw little activity on the development of data access and re-use policy frameworks, with some

---

[11] For a more detailed examination of the development of Australian policy on access to public sector information, see A. Fitzgerald (2009). *Open Access Policies, Practices and Licensing: A review of the literature in Australia and selected jurisdictions*, http://eprints.qut.edu.au/28026/ (accessed on 18/10/2011); B. Fitzgerald (ed.) (2010). *Access to Public Sector Information: Law, Technology and Policy*. Sydney: Sydney University Press, http://eprints.qut.edu.au/34085/ (accessed on 18/10/2011); A. Fitzgerald, B. Fitzgerald and N. Hooper (2010). "Enabling open access to public sector information with Creative Commons Licences: the Australian experience". In Fitzgerald, B. (ed.). *Access to Public Sector Information: Law, Technology and Policy.* Sydney: Sydney University Press, http://eprints.qut.edu.au/29773/ (accessed on 18/10/2011); A. Fitzgerald (2010). "Open access and public sector information: policy developments in Australia and key jurisdictions". In B. Fitzgerald (ed.). *Access to Public Sector Information: Law, Technology and Policy.* Sydney: Sydney University Press, http://eprints.qut.edu.au/31024/ (accessed on 18/10/2011). The various positions adopted by the federal and State governments in relation to access to and pricing of spatial data were examined and analysed in the report produced by PricewaterhouseCoopers for ANZLIC (Price Waterhouse Coopers (2010). *Economic Assessment of Spatial Data Pricing and Access – Stage 1 Report: Principles, Issues and Alternative Models*, http://www.anzlic.org.au/Latest+News/275.aspx (accessed on 18/10/2011)).

[12] R. Buckeridge and T. Cutler (1994). *Commerce in Content: Building Australia's International Future in Interactive Multimedia Markets. Report for the Department of Industry, Science and Technology, CSIRO and the Broadband Services Expert Group*, http://www.nla.gov.au/misc/cutler/cutlercp.html (accessed on 18/10/2011).

[13] Ibid.

[14] Ibid.

[15] The National Library of Australia's submission to the Australian Science and Technology Council (ASTEC) is available online but The Networked Nation report or the government's response to it is not. See Library of Australia (1994). *Submission to Australian Science & Technology Council Inquiry into Research Data Networks*, http://trove.nla.gov.au/work/13936350 (accessed on 18/10/2011).

[16] Cutler & Company (2008). *Venturous Australia - Building Strength in Innovation. Report on the Review of the National Innovation System, prepared for the Australian Government Department of Innovation, Industry, Science and Research*, http://www.innovation.gov.au/Innovation/Policy/Pages/ReviewoftheNationalInnovationSystem.aspx (accessed on 18/10/2011).

notable exceptions. Of particular note is the *Spatial Data Access and Pricing Policy*[17] adopted by the federal government in 2001, which remains the most comprehensive data access and re-use policy developed in Australia to date and the only one applying on a national basis.

The last few years have seen a revival of interest in the potential of PSI, and an examination of the issue in a series of inquiries commencing with the Cutler review of the national innovation system in 2008. The Cutler review's green paper report *Venturous Australia – Building Strength in Innovation* (*Venturous Australia*) recommended the development of a 'National Information Strategy to optimise the flow of information in the Australian economy'[18], the aims of which would include 'maximis[ing] the flow of government generated information, research, and content for the benefit of users (including private sector resellers of information)'[19]. The federal government's response to the Venturous Australia report was generally supportive of its recommendations on access to PSI, accepting the need to build on initiatives already underway[20] and 'to develop a more coordinated approach to Commonwealth information management, innovation and engagement'[21]. At the same time, Australia endorsed the *Seoul Declaration on the Future of the Internet Economy*[22] at the OECD Ministerial Meeting in June 2008. The Seoul Declaration incorporates key principles from the OECD's *Recommendation of the Council for Enhanced Access and More Effective Use of Public Sector Information*[23] (OECD PSI Recommendation). In the Declaration, the OECD Member States undertake to foster creativity in the development, use and application of the Internet, through policies that:

- Maintain an open environment that supports the free flow of information, research, innovation, entrepreneurship and business transformation.
- Make public sector information and content, including scientific data, and works of cultural heritage more widely accessible in digital format[24].

The Seoul Declaration and the OECD PSI Recommendation's principles (addressing matters such as openness, access, re-use conditions, copyright and pricing) have had considerable influence on thinking about PSI in Australia. Although not legally binding, OECD Recommendations have the status of legal instruments that describe standards or

---

[17] See Commonwealth Interdepartmental Committee on Spatial Data Access and Pricing (2001). *A Proposal for a Commonwealth Spatial Data Access and Pricing Policy*, http://spatial.gov.au/sites/default/files/legacy/osdm.gov.au/OSDM/Policies%2band%2bGuidelines/Spatial%2bData%2bAccess%2band%2bPricing/default.html (accessed on 18/10/2011).

[18] Cutler & Company (2008). *Venturous Australia - Building Strength in Innovation. Report on the Review of the National Innovation System, prepared for the Australian Government Department of Innovation, Industry, Science and Research*, http://www.innovation.gov.au/Innovation/Policy/Pages/ReviewoftheNationalInnovationSystem.aspx (accessed on 18/10/2011).

[19] Ibid.

[20] E.g. in the Australian Bureau of Statistics, Geoscience Australia and the Bureau of Meteorology.

[21] Department of Innovation, Industry, Science and Research (2009). *Powering Ideas: An Innovation Agenda for the 21st Century*, http://www.innovation.gov.au/innovation/policy/pages/PoweringIdeas.aspx (accessed on 18/10/2011).

[22] OECD (2008). The Seoul Declaration for the Future of the Internet Economy, http://www.oecd.org/dataoecd/49/28/40839436.pdf (accessed on 18/10/2011).

[23] OECD (2008). Recommendation of the Council for Enhanced Access and More Effective Use of Public Sector Information, http://www.oecd.org/dataoecd/0/27/40826024.pdf (accessed on 18/10/2011). The recommendation forms part of the annexes of the Seoul Declaration. See OECD (2008). *Shaping Policies for the Future of the Internet Economy, Annexes*, http://www.oecd.org/dataoecd/1/28/40821729.pdf (accessed on 18/10/2011). The Recommendation of the Council for Enhanced Access and More Effective Use of Public Sector Information is Annex F.

[24] OECD (2008). The Seoul Declaration for the Future of the Internet Economy, http://www.oecd.org/dataoecd/49/28/40839436.pdf (accessed on 18/10/2011).

objectives that OECD member countries are expected to implement, and are regarded as having great moral force.

The insights and proposals in Venturous Australia and the Seoul Declaration on the Future of the Internet Economy were soon informing other inquiries into the role of information flows in innovation and the potential for government information to be utilised to better effect. The report of the Victorian Parliament's Economic Development and Infrastructure Committee, *Inquiry into Improving Access to Victorian Public Sector Information and Data*, was tabled in the Victorian Parliament in June 2009[25]. The Committee had been asked to report on the benefits and costs of maximising access to and use of PSI for commercial and non-commercial purposes and to consider how flexible licensing arrangements would facilitate re-use of PSI. The key economic recommendation in the report was that the Victorian Government establishes a comprehensive Information Management Framework, with open access to PSI at no or marginal cost as the default position, and the development of specific guidelines to deliver that policy outcome[26]. The Committee's view was that the economic and social benefits arising from the release of Victorian government information at no cost would far outweigh the benefits gained from treating it as a commodity[27]. Next, the 2009 report of the federal government's Department of Broadband, Communications and the Digital Economy, *Australia's Digital Economy: Future Directions*, recognised 'the digital economy and innovation benefits generated by open access to public sector information, subject to issues such as privacy, national security and confidentiality'[28]. Enabling open access to government materials was seen as a means of promoting innovation in both the public and private sectors[29].

Progress towards the formation of an information strategy was boosted by the work of the Government 2.0 Taskforce (headed by Dr Nicholas Gruen) in 2009. The Taskforce's central recommendation was that a declaration of open government should be made at the highest level, stating that:

- Using technology to increase citizen engagement and collaboration in making policy and providing services will help achieve a more consultative, participatory and transparent government;
- PSI is a national resource and that releasing as much of it on as permissive terms as possible will maximise its economic and social value to Australians and reinforce its contribution to a healthy democracy;

[25] Victorian Parliament Economic Development and Infrastructure Committee (2009). *Inquiry into Improving Access to Victorian Public Sector Information and Data*, http://www.parliament.vic.gov.au/images/stories/committees/edic/access_to_PSI/EDIC_ACCESS_TO_PSI_REPORT_2009.pdf (accessed on 18/10/2011). The main recommendations are summarised in the accompanying media release: Economic Development and Infrastructure Committee (2009). *21st Century Approach to Government Information: Committee calls for improved access to government information*, http://www.parliament.vic.gov.au/edic/inquiries/article/1019 (accessed on 18/10/2011).

[26] Ibid.

[27] Ibid.

[28] Department of Broadband, Communications and the Digital Economy (2009). *Australia's Digital Economy: Future Directions*, http://www.dbcde.gov.au/digital_economy/what_is_the_digital_economy/ australias_digital_economy_future_directions/final_report (accessed on 18/10/2011).

[29] Ibid.

- Online engagement by public servants, involving robust professional discussion as part of their duties or as private citizens, benefits their agencies, their professional development, those with whom they are engaged and the general public[30].

The Government 2.0 Taskforce also recommended the adoption of a general policy that PSI should be 'open, accessible and reusable' and should, by default, be:

- Free;
- Based on open standards;
- Easily discoverable;
- Understandable;
- Machine-readable;
- Freely reusable and transformable[31].

These recommendations were accepted outright by the Australian government[32]. The government's response expressly stated its commitment to the principles of openness and transparency and its intention to affirm those principles through a Declaration of Open Government[33]. On 16 July 2010 the (then) Minister for Finance and Deregulation, Lindsay Tanner, announced the *Declaration of Open Government*, which acknowledges the importance of better access to and re-use of government-held information and the innovative use of technology in achieving a more open, participatory and transparent democracy. The Declaration outlines three key principles underlying the government's support for openness and transparency in government:

- Informing: strengthening citizen's rights of access to information, establishing a pro-disclosure culture across Australian Government agencies including through online innovation, and making government information more accessible and usable;
- Engaging: collaborating with citizens on policy and service delivery to enhance the processes of government and improve the outcomes sought; and
- Participating: making government more consultative and participative[34].

Launching the Declaration of Open Government, Mr Tanner said that:

> "[I]t acknowledges that the internet holds a crucial role in realising a more open and transparent form of government in this country. The Declaration is about making more government information available to the public online, and encouraging reuse of that information in new, valuable and potentially unexpected ways"[35].

---

[30] Government 2.0 Taskforce (2009). Engage: *Getting on with Government 2.0*, http://www.finance.gov.au/publications/gov20taskforcereport/doc/Government20TaskforceReport.pdf (accessed on 18/10/2011).

[31] Ibid.

[32] Australian Government (2010). *Government Response to the Report of the Government 2.0 Taskforce*, http://www.finance.gov.au/publications/govresponse20report/doc/Government-Response-to-Gov-2-0-Report.pdf (accessed on 18/10/2011).

[33] Ibid.

[34] L. Tanner (2010). *Declaration of Open Government*, http://agimo.govspace.gov.au/2010/07/16/declaration-of-open-government/ (accessed on 18/10/2011).

[35] Ibid.

The default position of openness of PSI is reiterated in the 2011 *Principles on open public sector information* developed by the Office of the Australian Information Commissioner through a process of public consultation[36]. Principle 1 states that PSI held by federal government agencies 'is a valuable national resource' which should be 'open to public access' if there is no legal need to protect it[37].

From 'freedom of information' to the 'right to information'

Much work has been done since the mid 2000s on overhauling and strengthening the FOI schemes that were first introduced in Australia during the 1980s and 1990s[38]. The current phase of FOI reform activity commenced in 2007 when the Queensland government appointed the FOI Independent Review Panel (Review Panel), headed by Dr David Solomon, to review that State's FOI regime. Central to this phase of FOI reform is a shift from a 'pull' model under which PSI is only disclosed in response to specific requests, to a 'push' model whereby PSI is routinely and proactively made available for access and re-use.

Following publication of the Review Panel's report *The Right to Information*[39] in June 2008, the Queensland government responded by enacting the *Right to Information Act* 2009 (Qld) to give effect to its recommendations. The primary object of this act is to 'give a right of access to information in the government's possession or under the government's control unless, on balance, it is contrary to the public interest' to provide access[40]. By enacting the Right to Information Act 2009 (Qld) the Queensland government sought to emphasise and promote the right to access government information[41], with PSI being released administratively as a matter of course, unless there are good reasons for withholding it, such that formal applications for release should only be necessary as a last resort[42]. Government agencies are required to publish a publication scheme setting out the classes of information available and the terms on which it is made available, including any charges[43].

The New South Wales government also overhauled its FOI laws in 2009, enacting the *Government Information (Public Access) Act* 2009 (GIPA)[44] and the *Government Information (Information Commissioner) Act* 2009. The GIPA authorises and encourages the proactive public release of government information, establishing a 'presumption in favour of disclosure of government information unless there is an overriding public interest against disclosure'[45].

---

[36] Australian Government. Office of the Australian Information Commissioner (2011). *Principles on open public sector information*, http://www.oaic.gov.au/publications/agency_resources/principles_on_psi_short.pdf (accessed on 18/10/2011). The Principles are intended for use in monitoring compliance by federal government agencies with the Freedom of Information Act 1982 (Cth) but are not otherwise binding.

[37] Ibid.

[38] Freedom of Information Act 1982 (Cth), Freedom of Information Act 1989 (ACT), Freedom of Information Act 1991 (Tas.), Freedom of Information Act 1989 (N.S.W.), Freedom of Information Act 1992 (Qld), Freedom of Information Act 1991 (S.A.), Freedom of Information Act 1982 (Vic.), Freedom of Information Act 1992 (W.A.).

[39] D. Solomon, S. Webbe and D. McGann (2008). *The Right to Information: Reviewing Queensland's Freedom of Information Act,* http://www.qld.gov.au/about/rights-accountability/right-to-information/assets/solomon-report.pdf (accessed on 22/5/2010).

[40] Right to Information Act 2009 (Qld), s 3(1). See also the Preamble the Act, paragraph 3.

[41] Right to Information Act 2009 (Qld), s 23.

[42] Right to Information Act 2009 (Qld), Preamble, paragraph 2.

[43] Right to Information Act 2009 (Qld), s 21.

[44] The Government Information (Public Access) Act 2009 (NSW) entered into force on 1 July 2010.

[45] Government Information (Public Access) Act 2009 (NSW), s 5.

The GIPA provides for release of government information in four sets of circumstances:[46]

- Mandatory disclosure of open access information: Agencies must publish certain information on their website, free of charge. 'Open access information', which must be published, consists of:
  - information about an agency's policies, structure and functions;
  - documents tabled in Parliament on the agency's behalf;
  - policy documents;
  - general details of unpublished open access information;
  - publication guide;
  - disclosure log;
  - register of government contracts.
- Proactive release: Agencies are encouraged to proactively release as much government information as possible, in an appropriate manner and free of charge or at the lowest reasonable cost. Any government information can be proactively released, unless there is an overriding public interest against its disclosure.
- Informal release: Agencies are encouraged to release information in response to a request without the need for a formal application, unless there are good reasons to require one.
- Formal access application: In limited circumstances, access to information will require a formal access application. There is a right to access information this way, unless the Act provides a reason to withhold the information.

Next to New South Wales and Queensland, on 13 May 2010, the federal Parliament enacted legislation that provides a new legal foundation for its policy of promoting greater openness and transparency in government: the *Freedom of Information Amendment (Reform) Act* 2010 and the *Information Commissioner Act* 2010. The amendments to the FOI legislation – the most significant refurbishment since the scheme was first introduced in 1982 – establish a default position of open access to PSI. The FOI Reform Act established a statutory framework for a pro-active publication scheme for PSI and introduced a single public interest test which favours disclosure. Following the changes in the FOI legislation, the *Principles on open public sector information* issued by the Office of the Australian Information Commissioner in May 2011 encourage federal agencies to apply a presumption of openness to the PSI they hold, to adopt a proactive publication stance to enhance public access and to use information technology to disseminate their information[47].

[46] A similar model has been adopted in the Tasmanian Right to Information Act 2009, which establishes the following four categories of disclosure of information: a required disclosure; a routine disclosure; an active disclosure; and an assessed disclosure (section 12(2)). The principal officer of a public authority is to ensure that adequate processes are in place to ensure that there is appropriate active disclosure, routine disclosure or required disclosure of information: section 12(3). An 'assessed disclosure' is the method of disclosure of last resort: s 12(3).

[47] Australian Government. Office of the Australian Information Commissioner (2011). *Principles on open public sector information*, http://www.oaic.gov.au/publications/agency_resources/principles_on_psi_short.pdf (accessed on 18/10/2011).

Management of government copyright to support access and re-use

A vast range of materials produced by and for federal, State/Territory and local governments and their constituent departments and agencies – including spatial data – is subject to copyright. Under the *Copyright Act* 1968 (Cth), with a few exceptions government copyright is treated the same as copyright owned by non-government parties insofar as the range of protected materials and the exclusive proprietary rights attaching to them are concerned. However, the rationale for recognizing copyright in public sector materials and vesting ownership of copyright in governments is fundamentally different to the main rationales underpinning copyright generally. The central justification for recognizing Crown copyright is to ensure that government documents and materials created for public administrative purposes are disseminated in an accurate and reliable form. Consequently, the exclusive rights held by governments as copyright owners must be exercised in a manner consistent with this rationale. Hence, governments should exercise their exclusive rights to ensure that their copyright materials are made available for access and re-use, in accordance with any laws and policies relating to access to public sector materials. While copyright law vests copyright owners with extensive bundles of exclusive rights which can be exercised to prevent others making use of the copyright material, in the case of Crown copyright materials these rights should rarely be asserted by government to deviate from the general rule that Crown copyright materials will be available for 'full and free reproduction' by the community at large.

Ownership of copyright by government agencies is dealt with in Part VII of the Copyright Act 1968 (the 'Crown copyright' provisions)[48]. The principal provisions on which government copyright is based are sections 176-178. Sections 176 and 178 provide that the government owns copyright in literary, dramatic, musical and artistic works, sound recordings and films 'made by, or under the direction or control of the Commonwealth or a State'. Section 177 further provides that the government owns copyright in a literary, dramatic, musical or artistic work that is first published in Australia 'by, or under the direction or control of, the Commonwealth or a State'[49]. The operation of sections 176-178 can be displaced by an agreement between the government and the person who created the copyright material that copyright is to belong to that person or some other party specified in the agreement[50]. The effect of these provisions is that governments own copyright in a vast range of materials in hard copy and digital form.

As the Copyright Act 1968 does not generally differentiate between the rights of government as copyright owner and the rights of private parties who own copyright, governments enjoy the same range of exclusive rights in their copyright materials as

---

[48] See A. Fitzgerald, B. Fitzgerald and N. Hooper (2010). "Enabling open access to public sector information with Creative Commons Licences: the Australian Experience". In Fitzgerald, B. (ed.). *Access to Public Sector Information: Law, Technology & Policy*. Sydney: Sydney University Press, http://eprints.qut.edu.au/29773/ (accessed on 18/10/2011); A. Fitzgerald and K. Pappalardo (2009). *Report to the Government 2.0 Taskforce: Project 4 – Copyright Law and Intellectual Property*, http://eprints.qut.edu.au/29416/ (accessed on 18/10/2011).

[49] Sections 176-178 are subject to any agreement between the Crown and the maker of the work or subject matter under which it is agreed that copyright is to belong to the author or maker or some other specified person (section 179).

[50] Copyright Act 1968, s179.

private sector copyright owners[51]. One of the few points of difference is that the duration of copyright of the government for materials within the scope of sections 176-178 is 50 years from the end of the calendar year in which the copyright item is first published or is made.[52]

Copyright endows right owners with extensive exclusive rights that can be exercised to control the use and re-use of their materials, including the right to:

- Reproduce;
- Publish;
- Publicly perform;
- Make an adaptation; and
- Communicate the copyright work to the public in electronic form (e.g. on a website or as a digital file)[53].

Other rights of copyright owners are the rights to ensure that electronic rights management information (ERMI) is not removed or altered and to prevent the circumvention of technological protection measures (TPM) they apply to their copyright materials to control access to or copying of it. ERMI is electronic information (including numbers or codes representing such information) which is either attached to or embodied in the copyright material, or appears in connection with a communication or the making available of the copyright material[54]. It typically includes information identifying the copyright work, its author or copyright owner, information indicating the terms and conditions on which the material can be used, or a statement that the use of the material is subject to terms or conditions. It is an infringement of the copyright owner's rights to remove or alter ERMI relating to a copyright work or other subject matter without the permission of the copyright owner or exclusive licensee, if the person doing the act knows or ought reasonably to have known that the removal or alteration would induce, enable, facilitate or conceal an infringement of copyright[55]. In certain circumstances the removal or altering of ERMI relating to a copyright work may be a criminal offence under the Copyright Act[56]. The provisions on the anti-circumvention of TPMs enable copyright owners to protect their materials by applying technical measures that control access to or copying of the work. It is an infringement to knowingly deal in devices designed to circumvent TPMs[57] and, where the TPM controls access to a copyright work, it is an infringement to knowingly circumvent the TPM[58].

[51] Section 182 specifically states that, apart from the provisions in Part VII of the Copyright Act 1968 (in ss 176-181) relating to the subsistence, duration and ownership of copyright, the provisions of Part III and Part IV of the Act apply.

[52] Copyright Act 1968, ss 180, 181.

[53] Copyright Act 1968, ss 31, 85-88.

[54] The main provisions dealing with ERMI are set out in Division 2A, Subdivision B of the Copyright Act 1968. Section 116D sets out the legal remedies (including an injunction or damages) available for the removal of and interference with ERMI.

[55] Copyright Act 1968, ss 116B-116D.

[56] Copyright Act 1968, ss 132AQ-132AS.

[57] Copyright Act 1968, s116AO(1).

[58] Copyright Act 1968, s 116AN(1).

Rationale for government copyright ownership

As mentioned earlier, the Copyright Act 1968 does not generally differentiate between public and private sector parties either with respect to the scope of materials in which copyright subsists or the exclusive rights that can be exercised in respect of them. However, it would be a mistake to assume that government copyright is exactly the same as copyright in non-government materials. Although statutory recognition of government copyright ownership was introduced into Australian copyright law in 1912 when the United Kingdom's *Copyright Act* 1911 was adopted in Australia, the rationale for copyright in government materials is quite different from the reasons why copyright in recognized in materials produced by individual authors and private sector organisations.

An obvious point of difference is that, since many government materials (e.g. reports, legislation, handbooks) are created in the ordinary course of activities by parliament, the courts and government agencies, the traditional justification of copyright as providing an incentive to produce and disseminate new information is much less relevant than for works produced by publishers with the expectation of a commercial return[59]. In fact, discussion of government copyright is strikingly lacking in the usual rationales for recognition of copyright, such as encouragement of innovation or enterprise for commercial benefit; reward of creative effort; maximizing commercial return through sale or licensing; and securing some advantage through the exercise of the exclusive rights.

The Copyright Law Review observed in the *Crown Copyright* report (2005) that works such as legislation and judgments 'will be produced regardless of financial incentives, and therefore the traditional justification for copyright ownership does not apply'. Continued recognition of government copyright is justified by the 'need to ensure the integrity and authenticity of official government publications'[60]. Ensuring that government materials are reproduced and distributed in an accurate and reliable form is a recurring theme in discussions about the existence and exercise of government copyright. Familiar phrases reappear: the need to ensure the authentic and accurate publication of documents, to be able to rely on the veracity and accuracy of government materials, and to indicate the status and authority of government materials.

Exercise of exclusive rights by governments

Although the same set of exclusive rights applies to the same materials, it was not intended that those rights would be exercised by governments in the same way as non-government copyright owners. Just as the rationale for government copyright ownership differs from that for private sector copyright, there is clear evidence that it was intended that government rights would be exercised primarily to ensure the distribution of government publications in a reliable form. Documents retrieved in archival research show that the concept of Crown copyright in United Kingdom and Australian law was, at least from the time it was first codified in statutory form, inextricably linked with what would now be termed 'open content' licensing practices. At the time the first Crown copyright provisions were enacted

---

[59] Copyright Law Review Committee (2005). *Crown Copyright*, http://www.ag.gov.au/Copyright/CopyrightLawReviewCommittee/Reports/Crowncopyright/Pages/default.aspx (accessed on 18/10/2011).
[60] Ibid.

in the United Kingdom and Australia, it was explained that the Crown's rights would be exercised to permit the 'full and free reproduction' and widespread dissemination of the great bulk of government copyright materials[61].

There has been an ongoing tension between, on the one hand, adopting an open access approach and, on the other hand, focusing on cost recovery or generating commercial returns or rents. Over the years, many government agencies have, in reliance on their rights as copyright owners, imposed restrictions on the use and re-use of their materials, often to preserve commercial arrangements with licensees or resellers of materials supplied by government[62]. A report prepared by PricewaterhouseCoopers for the Australia New Zealand Land Information Council (ANZLIC) in 2010 observes that, where spatial information is provided by government agencies at prices above marginal cost, restrictions are imposed on its distribution and re-use and royalties are payable for revenue obtained from the commercial use of the data[63]. For example, under the cost recovery default pricing approach for spatial data in Victoria, re-use is restricted and a royalty is payable on any commercial exploitation of data for new or value added products and services that generate a revenue stream[64]. More liberal licensing practices are typically found when spatial data is provided at marginal cost pricing. Recent developments have seen a reversal of licensing practices based on government copyright ownership and the introduction of a default position that any government material that is available for access should be able to be used and re-used, subject to proper attribution of its origin.

Implementation of CC licensing in the public sector

A striking feature of the Australian experience in moving towards greater accessibility and reusability of public sector materials generally – and spatial data specifically – is the extent to which government materials are being distributed under CC licences. The potential for the applicability of CC licences on government materials was first considered in a project initiated by the Queensland Spatial Information Council in 2004 to address long-standing dissatisfaction with the perceived limitations of the prevailing legal arrangements and practices for data access and sharing, both within government and between government and the private sector[65]. Having seen the potential relevance of CC licences, from 2005 work proceeded collaboratively between QUT's Law Faculty (managers of the Australian CC project and licences), the Queensland Government's Office of Economic and Statistical

[61] B. Atkinson (2007). *The True History of Copyright: The Australian Experience 1905 – 2005*. Sydney: Sydney University Press; B. Fitzgerald, A. Fitzgerald et al. (2011). *Internet and E-Commerce Law, Business and Policy*. Sydney: Thomson Reuters.

[62] For a detailed examination of the kinds of restrictive clauses imposed by Queensland government agencies when licensing spatial data, see Queensland Spatial Information Office (2006). *Government Information and Open Content Licensing: An Access and Use Strategy*, http://eprints.qut.edu.au/32117/1/Stage%202%20Final%20Report%20 PDF%20Format.pdf (accessed on 18/10/2011).

[63] Price Waterhouse Coopers (2010). *Economic Assessment of Spatial Data Pricing and Access – Stage 1 Report: Principles, Issues and Alternative Models*, http://www.anzlic.org.au/Latest+News/275.aspx (accessed on 18/10/2011).

[64] Ibid.

[65] Initial consideration of the applicability of CC licences to government copyright materials occurred in response to a request to examine this issue from Tim Barker, (then) Assistant Government Statistician and Director, Queensland Spatial Information Office, Office of Economic and Statistical Research, Queensland Treasury, Graham McColm, Principal Advisor, Department of Natural Resources and Water, Queensland and Rob Bischoff.

Research[66] and the Department of Natural Resources and Water (now Department of Environment and Resource Management). This project, which developed the CC-based Government Information Licensing Framework (GILF) and investigated legal, policy and economic[67] aspects of access to PSI, provided input into the various reviews of this area and led the way towards the adoption of CC licensing in the government sector in Australia.

When CC launched in Australia in 2005, the advantages of open content licensing in the education sector were already appreciated[68] and there was a dawning awareness that a similar approach could play a role in facilitating access to government copyright materials[69]. Submissions to the Copyright Law Review Committee's inquiry into Crown Copyright (2004-2005) urged the Committee to go beyond simply considering whether an improved access regime required abolition of Crown copyright (entirely or in part), and to examine whether the objective could be attained by the adoption of more flexible, open, standardised licensing practices[70]. Immediately upon the launch of the Australian CC licences in 2005, spatial information professionals in the Queensland public sector who had been seeking to rationalise and streamline the licensing arrangements for spatial and statistical information turned their attention to the potential application of the new suite of licences in this context[71].

From the outset, the principal focus of the project was the development of a standardised information licensing model for PSI that could be recommended for use with all kinds of government copyright materials to enable enhanced, seamless, on-demand access. The project sought to actively counteract the tendency towards licence proliferation, taking a strong view on the importance of a standardised approach and seeking to avoid the complexities that would inevitability re-occur if different government agencies and different jurisdictions insisted on developing individual licensing templates for access to and re-use of their copyright content. Significantly, the project did not directly address information policy or the development of principles for PSI access and re-use.

---

[66] The Office of Economic and Statistical Research is in Queensland Treasury and throughout this time was managed by Dr Peter Crossman.

[67] J. Cook (2010). *A summary view of government cost recovery policies in Australia and New Zealand relating to the supply of public sector information*, http://eprints.qut.edu.au/31609/ (accessed on 18/10/2011); Cook, J. (2009). Economic issues in funding and supplying public sector information, http://eprints.qut.edu.au/27832/ (accessed on 18/10/2011).

[68] The use of standardised licensing in the education sector in Australia had been pioneered by AEShareNET. See further, B. Fitzgerald et al. (2006). *Creating a Legal Framework for Copyright Management of Open Access within the Australia Academic and Research Sector*, www.oaklaw.qut.edu.au/files/LawReport/OAK_Law_Report_v1.pdf (accessed on 18/10/2011); B. Fitzgerald (2006). "Open Content Licensing (OCL) for Open Educational Resources". Paper presented at the OECD Expert Meeting on Open Educational Resources, Malmo, 6 - 7 February.

[69] Digital Content Industry Strategic Industry Leaders Group (2006). *Unlocking the Potential: Digital Content Industry Action Agenda Report*, http://www.archive.dcita.gov.au/2007/12/unlocking_the_potential_digital_content_industry_action_agenda_report (accessed on 18/10/2011); S. Cunningham et al. (2007). "Why Governments and Public Institutions Need to Understand Open Content Licensing". In Fitzgerald, B., Coates, J. and Lewis, S. (ed.). *Open Content Licensing: Cultivating the Creative Commons*. Sydney: Sydney University Press, 74-92.

[70] See http://www.ag.gov.au/Copyright/CopyrightLawReviewCommittee/Reports/Crowncopyright/Pages/CrownCopyrightSubmissions.aspx (accessed on 18/10/2011).

[71] For several years, these officers had been investigating ways of improving the flow of spatial information within the Queensland Government, and between the State and other levels of government and the private sector. They had viewed a video presentation by Professor Lawrence Lessig delivered at an event at QUT in 2004 to mark the launch of Creative Commons in Australia and immediately grasped the potential for CC licences to be applied towards achieving their objective of reducing impediments to the flow of spatial information.

The 2006 report *Government Information and Open Content Licensing: An Access and Use Strategy*[72] ('the Stage 2 report') identified the need for clear and succinct guiding principles for access, re-use and pricing and concluded that CC licences were the most appropriate available open licences for PSI. It supported the introduction of a simplified system of open content licensing for the majority of the information made publicly available by the Queensland government and recommended:

- That the Queensland Government established a policy position that, while ensuring that confidential, security classified and private information collected and held by government continues to be appropriately protected, enables greater use and re-use of other publicly available government data and facilitates data sharing arrangements.
- That the CC open content licensing model be adopted by the Queensland Government to enable greater use of publicly available government data and to support data sharing arrangements[73].

By focusing attention on the importance of removing barriers to access to and re-use of PSI caused by inadequate or inappropriate licensing practices, the project's findings and recommendations about the use of CC licences influenced reviews of information access policies and practices by the federal government[74], other State governments[75], the New Zealand Government[76] and the United Kingdom government[77]. Having firstly attracted the interest of the Queensland government's Chief Information Office, the project was ultimately reviewed and supported in principle by the Cross-Jurisdictional Chief Information Officers Committee and received the endorsement of the Ministerial Online and Communications Council in 2007.

By the time of the Cutler review of the national innovation system in 2008, the idea of using CC licences on public sector materials gained broader acceptance throughout Australia. The Venturous Australia report recommended that 'Australian governments should adopt international standards of open publishing as far as possible [and that] material released for public information by Australian governments should be released under a creative commons licence'[78]. At the State level, corresponding recommendations were made by the Victorian Parliament's Economic Development and Infrastructure Committee, which

[72] Queensland Spatial Information Office (2006). *Government Information and Open Content Licensing: An Access and Use Strategy*, http://eprints.qut.edu.au/32117/1/Stage%202%20Final%20Report%20PDF%20Format.pdf (accessed on 18/10/2011).

[73] Ibid.

[74] Tam, S. (2009). "Informing the Nation – Open Access to Statistical Information in Australia". Paper presented to the United Nations Economic Commission for Europe Work Session on the Communication and Dissemination of Statistics, www.unece.org/stats/documents/ece/ces/ge.45/2009/wp.11.e.pdf (accessed on 18/10/2011).

[75] Victorian Parliament Economic Development and Infrastructure Committee (2009). *Inquiry into Improving Access to Victorian Public Sector Information and Data*, http://www.parliament.vic.gov.au/images/stories/committees/edic/access_to_PSI/EDIC_ACCESS_TO_PSI_REPORT_2009.pdf (accessed on 18/10/2011).

[76] The New Zealand Government Open Access and Licensing Framework (NZGOAL) uses the New Zealand Creative Commons licences. See http://www.e.govt.nz/guidance-and-resources/information-and-data/nzgoal (accessed on 18/10/2011) and http://nzgoal.info/ (accessed on 18/10/2011).

[77] See Power of Information Taskforce (2009). *Power of Information Task Force Report*, http://webarchive.nationalarchives.gov.uk/20100413152047/http://poit.cabinetoffice.gov.uk/poit/wp-content/uploads/2009/03/poit-report-final-pdf.pdf (accessed on 18/10/2011).

[78] Cutler & Company (2008). *Venturous Australia - Building Strength in Innovation. Report on the Review of the National Innovation System, prepared for the Australian Government Department of Innovation, Industry, Science and Research*, http://www.innovation.gov.au/Innovation/Policy/Pages/ReviewoftheNationalInnovationSystem.aspx (accessed on 18/10/2011).

recommended the adoption of the CC licences as the default licences for the proposed Information Management Framework[79].

The Government 2.0 Taskforce recommended in 2009 that '[c]onsistent with the need for free and open re-use and adaptation PSI should be licensed under the Creative Commons BY standard as the default'[80]. More restrictive licensing conditions should be applied only in special circumstances, in accordance with general guidance or specific advice provided by the Australian Information Commissioner[81]. The Taskforce proposed that the CC BY licence should be the default licence applied to PSI for which the government owns copyright, as well as PSI containing third party material. In the latter case it will of course be necessary to negotiate with the copyright owners in order to obtain permission to distribute the material under a CC licence[82]. Existing public sector materials that have already been published (such as government reports, legislation and publicly accessible records) could be relicensed under a CC BY licence, either on the initiative of the government or on request by individuals or organisations[83]. Each of these recommendations was accepted[84] and is being implemented by the federal government. To ensure that the federal government's copyright management practices do not impede the application of CC BY as the default licensing option, the government directed the Attorney General's Department to review its IP Guidelines[85].

In October 2010 the Attorney General's Department released an updated *Statement of IP Principles for Australian Government Agencies*, formally endorsing the Government 2.0 Taskforce's recommendation that agencies should license their PSI under a CC BY licence as the default position and only adopt more restrictive licences after a process of 'due diligence and on a case-by-case basis'[86]. The IP Principles state:

> "11.(b) Consistent with the need for free and open re-use and adaptation, public sector information should be licensed by agencies under the Creative Commons BY standard as the default.
>
> An agency's starting position when determining how to license its public sector information should be to consider Creative Commons licences (http://creativecommons.org.au/) or other open content licences.
>
> Agencies should license their public sector information under a Creative Commons licence or other open content licence following a process of due diligence and on a case-by-case basis.

[79] Victorian Parliament Economic Development and Infrastructure Committee (2009). *Inquiry into Improving Access to Victorian Public Sector Information and Data*, http://www.parliament.vic.gov.au/images/stories/committees/edic/access_to_PSI/EDIC_ACCESS_TO_PSI_REPORT_2009.pdf (accessed on 18/10/2011).

[80] Government 2.0 Taskforce (2009). *Engage: Getting on with Government 2.0*, http://www.finance.gov.au/publications/gov20taskforcereport/doc/Government20TaskforceReport.pdf (accessed on 18/10/2011).

[81] Ibid.

[82] Ibid.

[83] Ibid.

[84] Australian Government (2010). *Government Response to the Report of the Government 2.0 Taskforce*, http://www.finance.gov.au/publications/govresponse20report/doc/Government-Response-to-Gov-2-0-Report.pdf (accessed on 18/10/2011).

[85] Ibid.

[86] Attorney-General's Department (2010). *Statement of IP Principles for Australian Government Agencies*, http://www.ag.gov.au/Documents/Statement%20of%20IP%20Principles%20for%20Australian%20Government%20Agencies2.pdf (accessed on 18/10/2011).

> Before releasing public sector information, for which the Commonwealth is not the sole copyright owner, under a Creative Commons BY standard or another open content licence, an agency may need to negotiate with any other copyright owners of the material"[87].

In line with these principles, the *Principles on open public sector information* published by the Office of the Australian Information Commissioner in May 2011 also acknowledge that the 'economic and social value of PSI is enhanced when it is made available for re-use on open licensing terms' and specify that the default licence is CC BY[88].

Another of the Government 2.0 Taskforce's recommendations was that government copyright works held in official archives should be automatically licensed under a CC BY licence immediately upon becoming available for access by the public under the provisions of the *Archives Act* 1983 (Cth)[89]. The federal government accepted that, when Commonwealth records become available for public access under the Archives Act 1983, they should be automatically licensed under an appropriate open attribution licence, subject to the proviso that 'the selection and use of an appropriate open attribution licence will remain the responsibility of agencies on a case-by-case basis'[90]. In making the decision about which open attribution licence to apply, government agencies can use the licensing tool in GILF, which assists in the selection of CC licences[91]. The government's response to this recommendation has also been included in the revised Statement of IP Principles for Australian Government Agencies, which states that '[a]t the time at which Commonwealth records become available for public access under the Archives Act 1983, public sector information covered by Crown copyright should be automatically licensed under an appropriate open content licence'[92].

Importantly, in accepting the Government 2.0 Taskforce's recommendation that CC BY should be the default licence applied to government materials, the federal government has adopted a much more liberal position than that envisaged in the Queensland GILF project. As the Government 2.0 Taskforce chair Dr Nic Gruen emphasized, CC BY is a permissive licence 'which allows complete freedom to reproduce, and remix subject only to the acknowledgement of the original source'[93]. However, the GILF licensing scheme encompassed the full suite of six CC licences (some of which contain additional conditions not found in the Attribution licence) as well as a specifically drafted template restrictive licence and model clauses, for use where further restrictions are to be imposed. The availability of this range of licences, particularly those constructed using the restrictive licence and model

---

87 Ibid.

88 Australian Government. Office of the Australian Information Commissioner (2011). *Principles on open public sector information*, http://www.oaic.gov.au/publications/agency_resources/principles_on_psi_short.pdf (accessed on 18/10/2011).

89 Government 2.0 Taskforce (2009). *Engage: Getting on with Government 2.0*, http://www.finance.gov.au/publications/gov20taskforcereport/doc/Government20TaskforceReport.pdf (accessed on 18/10/2011).

90 Australian Government (2010). *Government Response to the Report of the Government 2.0 Taskforce*, http://www.finance.gov.au/publications/govresponse20report/doc/Government-Response-to-Gov-2-0-Report.pdf (accessed on 18/10/2011).

91 Ibid.

92 Attorney-General's Department (2010). *Statement of IP Principles for Australian Government Agencies*, http://www.ag.gov.au/Documents/Statement%20of%20IP%20Principles%20for%20Australian%20Government%20Agencies2.pdf (accessed on 18/10/2011).

93 Dr Nicholas Gruen (2010). Blog post at http://gov2.net.au/ (accessed on 18/10/2011).

clauses, meant that GILF enabled a much greater range of restrictions to be imposed on government materials than what was envisaged by the Government 2.0 Taskforce[94].

Even before the 2008 Venturous Australia Green Paper and the Government 2.0 Taskforce's inquiry, key federal custodians of spatial data whose activities were subject to the Spatial Data Access and Pricing Policy (2001)[95] had already taken steps to provide improved access to their PSI. From the mid 2000s, Geoscience Australia, the Australian Bureau of Statistics and the Bureau of Meteorology had progressively removed restrictions to the accessibility and reusability of their materials and all had adopted CC Attribution licenses in 2008, more than 12 months before the Government 2.0 Taskforce published its licensing recommendations. Under the *Spatial Data Access and Pricing Policy*, which is administered by the Office of Spatial Data Management[96], fundamental spatial datasets are to be provided free of charge over the Internet and at no more than the cost of delivery in the case of packaged products. Copyright in such data is maintained and re-use is permitted, subject to acknowledgement of copyright, and there are no restrictions on commercial use or value adding.

Geoscience Australia

Geoscience Australia (GA) was an early adopter of CC, being the first Australian government agency to apply the licences to its datasets in October 2008. Earlier that year, in response to requests from clients for easier access to GA's information products and clearer statements of the terms of use and re-use, GA undertook an analysis and internal trial of CC licences on a representative sample of its datasets to ascertain whether open content licensing would meet the organisation's desired operational outcomes. Following successful completion of the CC licensing trial, the Spatial Data Management Group, which had formed a Working Group to address the use of CC licensing, endorsed the use of CC licences for spatial data[97]. GA's Moderate Resolution Imaging Spectroradiometer, the Australian Atlas of Mineral Resources[98], the GeoMAP 250K dataset, digitised Bureau of Mineral Resources records and educational material about tsunami were among the first materials distributed under CC licences. In announcing its decision to apply CC licences to key mapping and other information products, GA emphasised that the use of the 'easy to understand, royalty-free,

---

[94] A similar more restrictive position was also accepted by the Victorian Parliament's Economic Development and Infrastructure Committee, which recommended that CC licences should be the default licences for the proposed Information Management Framework (recommendation 14). However, the committee also recommended (recommendation 15) that the Victorian government should adopt a hybrid licensing model for PSI, comprising the Creative Commons licences and a tailored suite of licences for restricted materials. See Victorian Parliament Economic Development and Infrastructure Committee (2009). *Inquiry into Improving Access to Victorian Public Sector Information and Data*, http://www.parliament.vic.gov.au/images/stories/committees/edic/access_to_PSI/EDIC_ACCESS_TO_PSI_REPORT_2009.pdf (accessed on 18/10/2011).

[95] See http://spatial.gov.au/sites/default/files/legacy/osdm.gov.au/OSDM/Policies%2band%2bGuidelines/Spatial%2bData%2bAccess%2band%2bPricing/default.html (accessed on 18/10/2011).

[96] Commonwealth Interdepartmental Committee on Spatial Data Access and Pricing (2001). *A Proposal for a Commonwealth Spatial Data Access and Pricing Policy*, http://spatial.gov.au/sites/default/files/legacy/osdm.gov.au/OSDM/Policies%2band%2bGuidelines/Spatial%2bData%2bAccess%2band%2bPricing/default.html (accessed on 18/10/2011).

[97] Australian Government. Office of Spatial Data Management (2010). *Australian Government Spatial Data Activities 2009: A report of the Spatial Data Management Group*, http://spatial.gov.au/sites/default/files/legacy/osdm.gov.au/SDMG-AR%3d2009c945.pdf?ID=1043 (accessed on 18/10/2011).

[98] See the Atlas of Mineral Resources, Mines and Processing Centres (the 'Australian Mines Atlas', http://www.australianminesatlas.gov.au (accessed on 18/10/2011).

modular, off the shelf [CC] licences' would make it easier for visitors to GA's website to use and access information. In November 2009, GA began licensing all the material on its own website and the OzCoasts website[99] (which it hosts) under the CC Attribution 2.5 Australia licence (CC BY)[100]. Following the launch of version 3.0 of the Australian CC licences in 2010, GA's website was updated to state that, unless otherwise noted, all GA material on the site is licensed under the CC Attribution 3.0 Australia licence (CC BY)[101]. During 2010, GA began releasing processed satellite data that shows change in tropical forest cover in South East Asia and is used to calculate the rate and impact of climate change, under CC licences. In December 2010, a few weeks before the severe flooding throughout the eastern States, GA released digital elevation data, which is available for free download under a CC BY licence on the beta National Elevation Data Framework portal[102].

Australian Bureau of Statistics

In November 2005, the Australian Bureau of Statistics (ABS) abandoned the restrictive licensing practices it had previously applied, which had seen it charge fees for access to data and prohibit or restrict commercial downstream use by the licensee and/or others[103]. Since then the ABS has eliminated virtually all charges for data and restrictions on access and re-use, whether commercial or otherwise. Following the removal of charges, the number of hits and downloads of ABS publications increased dramatically. Downloads of statistical publications increased from around 90,000 in 2000-01 to over 960,000 in 2004-05 and then to 4.5 million in 2006-07, the first full year of free downloads[104]. The number of page views doubled from the end of 2005 to the end of 2007[105].

Nevertheless, even after the relaxation of licensing practices in 2005, any significant redistribution of information obtained from the ABS website still had to be licensed by the ABS. Although the ABS allowed broad use of its website content, often at no cost, the licensing process itself was seen as potentially acting as a barrier to those wishing to re-use

[99] See http://www.ozcoasts.gov.au/ (accessed on 18/10/2011).

[100] Note that some datasets such as MapConnect and GADDS could not be made available immediately under CC licences because the OSDM registration was embedded in these products.

[101] See the Copyright statement on the GA website, http://www.ga.gov.au/copyright.html (accessed on 18/10/2011).

[102] See http://nedf.ga.gov.au/geoportal/catalog/main/home.page (accessed on 18/10/2011). Note that the three second and nine second elevation data are available under CC BY licences but the one second data are presently only available to public sector agencies and their contractors, due to defence security reasons.

[103] Commencement of the use of the CC BY licence for ABS materials was accompanied by the following statement of purpose on the ABS website: "The Australian Bureau of Statistics (ABS) has introduced Creative Commons (CC) licensing for the bulk of the content on this website. This will lessen the restrictions on the use of free data from the website considerably by changing the copyright from "all rights reserved" to "some rights reserved". In effect, what the ABS is asking is only that it be acknowledged as the source of the data. People are free to reuse, build upon and distribute our data, even commercially. This makes a wealth of data readily available to the community, researchers and business, facilitating innovative research and development projects based on quality statistics, and promoting the wider use of statistics in the community, which is one of our core objectives"(http://www.abs.gov.au/websitedbs/D3310114.nsf/4a256353001af3ed4b2562bb00121564/8b2bdbc1d45a10b1ca25751d000d9b03?opendocument?utm_id=HPI (accessed on 18/10/2011).

[104] Price Waterhouse Coopers (2010). *Economic Assessment of Spatial Data Pricing and Access – Stage 1 Report: Principles, Issues and Alternative Models*, http://www.anzlic.org.au/Latest+News/275.aspx (accessed on 18/10/2011).

[105] S. Tam (2009). "Informing the Nation – Open Access to Statistical Information in Australia". Paper presented to the United Nations Economic Commission for Europe Work Session on the Communication and Dissemination of Statistics, www.unece.org/stats/documents/ece/ces/ge.45/2009/wp.11.e.pdf (accessed on 18/10/2011).

significant amounts of data. Following discussions with the open access community[106] and relevant government departments, in mid 2008 ABS decided to make information on its website freely and openly available for access and re-use. This decision was consistent with ABS's philosophy of access to information, as well as the Venturous Australia Green Paper, which urged Australian governments to adopt international standards of open publishing and the release of materials for public information under a CC licence[107].

On 18 December 2008, the ABS implemented CC licensing on its website and began making an extensive range of its statistical information products available online under a CC Attribution Australia licence. Implementation involved adding to the footer on every page of the ABS website an updated Copyright Statement, Disclaimer notice, CC symbols, information on how to attribute material sourced from the ABS website and a hyperlink to the CC licence. In effect, ABS makes its website material openly available, on condition that users acknowledge ABS as the source of the data.

Bureau of Meteorology

The *Water Act* 2007 (Cth) expanded the role of the Bureau of Meteorology (BoM) to include management of water information, with the establishment of the Australian Water Resources Information System (AWRIS)[108]. BoM is required to collect water information from a range of sources and to disseminate it for widespread re-use, including by publishing a National Water Account and periodic reports on water resource use and availability.

To ensure that water information provided to BoM under the Water Regulations 2008 can be widely re-used, BoM sought the support of the States and Territories for the adoption of a CC licensing framework for copyright-protected water datasets and databases. BoM recommends that each of the 260 data suppliers that are obliged to provide information to it under the *Water Regulations* 2008 should apply CC licences – and, specifically CC BY – to all the data it provides to AWRIS, so that it can be re-used by anyone on condition that the original data supplier is acknowledged[109]. Almost all data suppliers are now applying CC licences when providing their reports to BoM[110].

---

[106] Throughout this period, senior ABS officers consulted extensively with the Queensland Government-QUT project group that developed the CC-based Government Information Licensing Framework.

[107] For the background to the ABS's adoption of CC licences, see S. Tam (2009). "Informing the Nation – Open Access to Statistical Information in Australia". Paper presented to the United Nations Economic Commission for Europe Work Session on the Communication and Dissemination of Statistics, www.unece.org/stats/documents/ece/ces/ge.45/2009/wp.11.e.pdf (accessed on 18/10/2011); Cutler & Company (2008). *Venturous Australia - Building Strength in Innovation. Report on the Review of the National Innovation System, prepared for the Australian Government Department of Innovation, Industry, Science and Research*, http://www.innovation.gov.au/Innovation/Policy/Pages/ReviewoftheNationalInnovationSystem.aspx (accessed on 18/10/2011).

[108] See http://www.bom.gov.au/waterjobs/awris.htm (accessed on 18/10/2011).

[109] See http://www.bom.gov.au/water/regulations/dataLicensing/ccLicense.shtml (accessed on 18/10/2011). See also Australian Government. Office of Spatial Data Management (2010). *Australian Government Spatial Data Activities 2009: A report of the Spatial Data Management Group*, http://spatial.gov.au/sites/default/files/legacy/osdm.gov.au/SDMG-AR%3d2009c945.pdf?ID=1043 (accessed on 18/10/2011).

[110] By 2009 the 'vast majority of data suppliers' had agreed to supply their data under Creative Commons licences. See Australian Government. Office of Spatial Data Management (2010). *Australian Government Spatial Data Activities 2009: A report of the Spatial Data Management Group*, http://spatial.gov.au/sites/default/files/legacy/osdm.gov.au/SDMG-AR%3d2009c945.pdf?ID=1043 (accessed on 18/10/2011).

Australian Ocean Data Network

The Australian Ocean Data Network (AODN) is a program supported by the Australian Ocean Data Centre Joint Facility (AODCJF)[111], a joint venture between the six federal government agencies with primary responsibility for marine data. It was established with the intention that it would be Australia's digital oceans data commons, increasing the discoverability, availability, accessibility and reusability of marine data in the Australian marine region. The AODN does not generate any data itself but operates as a federated system in which data and metadata stored with the originating organisations are harvested by the AODN metadata store. It includes both publicly funded marine data contributed by the AODCJF agencies and data from private industry and not-for-profit organisations. A condition of participation in the AODN is that all data is contributed on an unencumbered basis so that it can be made freely accessible on the Internet at no charge to users. Data contributed to the AODN continues to be owned by each provider and the AODN's role is to publish information about it and provide direct access to the data. To ensure that the data will be publicly accessible and reusable, the AODN recommends that contributors provide it under an open content licence such as the CC Attribution licence[112].

Conclusion

The importance of ensuring the accessibility and reusability of government information – including spatial data – is now widely acknowledged in Australia. Reforms and revisions of Freedom of Information laws, which recognise enhanced rights to obtain access to information, have combined with the adoption of open licensing practices so that an extended range of government materials is now accessible, in a form that permits use and re-use. The federal government's commitment to greater accessibility and reusability is evidenced in the Declaration of Open Government (July 2010), which clearly states the objective of pursuing an open, transparent and participatory democracy, 'built on better access to and use of government held information, and sustained by the innovative use of technology'.

The acceptance by the federal government of the Government 2.0 Taskforce's recommendation that CC BY should be the default copyright licence for government materials provides a strong indication of its understanding that it is not sufficient to merely provide access but that materials should be available for use and re-use, with as few restrictions as possible. Governments across Australia have taken significant steps in developing the framework of laws, policies and information management practices required to give effect to this objective. Despite this progress, in a federation consisting of a national government, eight States and Territories and a myriad of local government authorities, there are real challenges in developing a truly national framework that enables information to flow within and among the various levels of government, as well as between government and private individuals and organisations. The need for intergovernmental cooperation to develop and implement a national information policy was recognised by the Government 2.0 Taskforce, which recommended that effort should be put into developing a national information

[111] See http://www.aodc.org.au (accessed on 18/10/2011).

[112] Australian Ocean Data Network (2011). *AODN Data Policy (version 1.0)*, http://imos.org.au/fileadmin/user_upload/shared/AODN/AODN_data_policy_v1.pdf (accessed on 18/10/2011).

policy, through the Council of Australian Governments on which each of the jurisdictions is represented[113]. The challenge ahead, in further developing the legal, policy and practice framework for access to and re-use of PSI, is to proceed on the basis that PSI is indeed a national resource and to develop strategies designed to ensure that its social and economic benefit is maximised.

[113] Government 2.0 Taskforce (2009). *Engage: Getting on with Government 2.0*, http://www.finance.gov.au/publications/gov20taskforcereport/doc/Government20TaskforceReport.pdf (accessed on 18/10/2011).

Policies Restricting Availability

Part 2

Summary of Location Privacy in the United States

Kevin D. Pomfret[1]

Introduction

Rapid developments in a number of different geospatial technologies make it relatively easy to collect, process, visualize, analyse, and distribute location and other types of geographic data. As a result, a variety of people, organisations and institutions can gain access to our current and past location. In addition, increasingly both government agencies and businesses will be able to identify us with a high degree of certainty based upon our movement patterns and other generally available information. Such collection and use of data on our location and movements is increasingly seen as endangering our privacy, as such data provides considerable information about our activities, encounters and behavioural patterns.

Although this topic is receiving a great deal of attention in the media and in Congress, the existing legal and policy framework in the United States is not very helpful with respect to what is privacy from a location standpoint. Legislation is outdated and there is very little applicable case law. As a result, policymakers, businesses and the general public are all struggling to define 'a reasonable expectation of privacy' from a location standpoint.

This chapter will discuss the existing state of law in the United States with respect to location privacy. It will begin with a discussion of the government's obligation to protect an individual's location privacy. This section will include a discussion on the Privacy Act of 1974 and the Freedom of Information Act, as well as the protections afforded by the Fourth Amendment of the U.S. Constitution. The chapter will then turn to the potential impact that efforts to protect location privacy may have on businesses. The chapter will conclude with a general analysis of issues to consider as laws and policies develop in the United States.

What is location privacy?

Defining privacy from a location standpoint is difficult for many reasons. First, it is a relatively new concept. Generally, in the past the only people who knew where we were at a particular point in time were the people we told and the people who saw us there. This began to change when 'sensors' became more common (e.g. cameras, recording devices, mobile phones, etc.), and more information on our location could be collected and saved. However, the effect on our privacy is becoming a much larger concern now that location information can be collected, distributed and analysed in near real time. This increasing awareness of the potential dangers of the distribution of our location information has resulted in a new push to regulate and define location privacy.

[1] Centre for Spatial Law and Policy, kevin@spatiallaw.com.

Another difficulty in defining location privacy is that it means different things to different people. One's sense of location privacy depends upon a number of factors, including one's culture, age and gender. For example, a 20 year old young man who has grown up playing on-line interactive video games is likely more comfortable with location-enabling technology than a 68 year old woman who lives in the same small town she grew up in. This variation in people's sense of privacy is very different from many other information items that are widely accepted as personal and/or sensitive, such as social security numbers and other unique government-issued identification, or one's financial information, religious affiliation or sexual orientation. While there is general agreement that these types of data are privacy-sensitive, this is not always the case for data on our location.

Moreover, there is a challenge in defining what location even means from a privacy standpoint. For example, a few bills in Congress attempt to regulate the collection, use and or transfer of 'precise geographic information'. However, does the term 'precise' refer to the accuracy of the location or spatial/temporal precision? In addition, should our location in public even be considered sensitive? For years we have disclosed our location, without concern, each time we went out in public. Should advancements in technology change that level of concern? For example, when I walk down the street at lunch time I do not feel my privacy is violated when a vendor offers me a sample of his fare in order to entice me to enter for lunch. Why should I mind if that same vendor were to send me a coupon via my smartphone as I approach? In neither instance have I given consent to use my geo-location information; however, from a privacy standpoint, the latter feels more threatening to some. The following sections examine whether the existing legal framework in the United States has an answer to these new concerns.

Obligation of government with respect to location privacy[2]

The Privacy Act of 1974

The *Privacy Act* of 1974[3] (the Privacy Act) governs the collection, maintenance, use, and dissemination of records on U.S. citizens held by U.S. federal government agencies that can be retrieved by the individual's name or unique identifier. The Privacy Act sets forth a number of requirements that federal agencies must follow with respect to any 'system of records' they keep on individuals. These requirements include:

- *Access.* An agency must give individuals access to any records it may have on him or her, including the right to copy and to make sure the records are complete and accurate.
- *Use and Sharing Conditions.* There are restrictions on how and to whom a government agency may disclose such records. Disclosure is limited to the consent of the individual or other specifically delineated conditions, including:
  - Use which is compatible for the purpose for which it was collected;
  - Statistical research;

[2] This discussion will focus on U.S. federal laws, not State and local laws. Although there are applicable State laws, many are similar to federal law.

[3] 5 USC §552a.

 - Law enforcement; and
 - Census purposes.
- *Minimization Requirements*. An agency is required to maintain in its records only the minimum amount of information 'relevant and necessary' to accomplish its purpose.
- *Penalties*. Failure to comply with these requirements can result in civil and criminal penalties.

The Privacy Act is over thirty years old. Technology with respect to the collection, use and distribution of geo-location information has evolved tremendously over that period. Therefore, it should be no surprise that the Information Security and Privacy Advisory Board, an advisory board to U.S. government agencies, found in 2009 that '[c]urrent government privacy laws do not provide sufficient limits on how geo-location information can be used' once it is collected by a government agency[4]. The Board's report recommended that the Office of Management and Budget (OMB) 'issue privacy guidelines on agency use of location information'[5].

Thus far there is no evidence to suggest that the federal government has taken any formal steps to address this issue. Moreover, OMB has not issued any guidelines with respect to how agencies should collect, use or store an individual's geo-location information. As a result, government agencies are still struggling with this important issue.

Freedom of Information Act

Another important piece of legislation in the United States relating to an individual's privacy is the *Freedom of Information Act*[6] (FOIA). The FOIA, which was signed into law in 1966, was established to promote an informed citizenry. The premise of the act is that records of government agencies should generally be made available to the public upon request at the cost of reproduction.

The FOIA has been amended and updated on a number of occasions. One of the major amendments, the *Electronic Freedom of Information Act Amendment* of 1996, requires all government agencies to make records available in electronic form, making it much easier to acquire, process, and analyse the requested information. As described further below, it also increases the privacy issues associated with geo-location information collected by the government.

The intent of the FOIA is to allow individuals access, upon request, to information that the government collects, uses or transmits. It does not apply to all types of information. There are currently nine exemptions to the disclosure requirements, including:

- Interest of national defence or foreign policy;
- Exemption by statute;
- Trade secrets or commercial or financial information;

[4] Information Security and Privacy Advisory Board (2009). *Towards a 21st Century Framework for Federal Government Privacy Policy*, http://www.securityprivacyandthelaw.com/uploads/file/ispab-report-may2009.pdf (accessed on 23/06/2011).
[5] Ibid.
[6] 5 U.S.C. § 552, as Amended By Public Law No. 104-231, 110 Stat. 3048.

- Law enforcement; and
- Disclosures that would result in *an unwarranted invasion of privacy* [emphasis added].

Improvements in technology making it relatively easy to combine and analyse a variety of data sets have raised new questions with regard to the FOIA. For example, at what point does releasing spatially-enabled data sets that may be combined with other data in such a way as to identify an individual, constitute an unwarranted invasion of that individual's privacy? In the court case *Multi-Ag Media v. Department of Agriculture*[7], Multi-Ag, a commercial vendor of agriculture data, made a FOIA request for geographic information system (GIS) records of farms receiving Department of Agriculture subsidies related to livestock. Although no names and addresses were to be given, the Department of Agriculture challenged having to release the data electronically, due to privacy concerns. In its decision, the court articulated the need to balance the value in releasing the information to the public with the potential privacy concerns[8]. The court found that there was a substantial privacy interest to be protected, because the information was in digital form and therefore could be aggregated with other information relatively easily to identify the individual farmers who received subsidies. Since livestock can be considered an asset, the court reasoned it was conceivable that someone could use that information to determine the financial worth of an individual receiving a subsidy. However, ultimately the court found that the public value in releasing the information was more important than the privacy interest to be protected and ordered the Department of Agriculture to release the information[9].

Similarly, in *Forest Guardians v. FEMA*[10], the Forest Guardians, a conservation group, made a FOIA request to the Federal Emergency Management Agency (FEMA) for electronic mapping files of homes insured under FEMA's National Insurance Flood Program. FEMA claimed these files were exempt under FOIA as they could possibly be aggregated with other available information to constitute an unwarranted invasion of privacy of individuals insured under the program. The court found that even though the electronic map files did not include names and addresses, FEMA did not have to disclose the information electronically as third parties could use the location of the homes to get names and addresses. Individuals might then be subject to unwanted solicitations, for example from insurance companies selling flood insurance[11].

These examples show that courts face a number of challenges when dealing with the issue of aggregation of government data with respect to location privacy. Technology, particularly geospatial technology, is rapidly changing. As a result, it is difficult for the courts to define a set of rules to follow. As described above, some courts have tried to examine the aggregation of digital datasets. However, given advancements in technology, any data set, digital or otherwise, will soon be easily aggregated with other data at little cost. As a result, differences between delivering data digitally or otherwise may have little value for FOIA purposes.

[7] 515 F3d 1224 (DC Cir. 2008).
[8] Ibid.
[9] Ibid.
[10] 410 F.3d 1214 (10th Cir. 2005).
[11] Ibid.

Fourth Amendment

Background

The Fourth Amendment to the U.S. Constitution is another important source of privacy protection in the United States. The Fourth Amendment provides that '[t]he right of the people to be secure in their persons, houses, papers, and effects, against unreasonable searches and seizures, shall not be violated'[12]. *U.S. v. Katz*[13] became the seminal case in which the protections of the Fourth Amendment were found to have created a right to privacy. The issue in this case concerned a phone call Katz had placed from a public phone booth. Law enforcement had intercepted the call without first obtaining a warrant. The government argued that they did not need a warrant because the call was made from a public place and therefore Katz had no expectation of privacy with respect to the call. However, the Court found that by entering a telephone booth and closing its door, Katz had a reasonable expectation that the conversation would not be heard. Therefore, the government needed to obtain a warrant before tapping phone lines to listen in on the call.

Justice Harlan's concurrent opinion in Katz set forth what has come to be known as the 'reasonable expectation of privacy' test[14]. The test consists of two prongs. First, it has to be determined whether an individual has exhibited an expectation of privacy. This is the objective component of the test. Second, it has to be assessed whether society is prepared to recognize such an expectation as reasonable. This is the subjective component. As will be discussed below, applying this test to location privacy has become a challenge for the courts, particularly given the pace of technological change. This has for instance been the case in disputes dealing with remote sensing and tracking.

Example: remote sensing

In 1986, the U.S. Supreme Court issued two important decisions with regard to the use of remote sensing and privacy. In *Dow Chemical Co. v. U.S*[15], the Court found that the Environmental Protection Agency did not violate the Fourth Amendment rights of Dow Chemical when it used a plane to obtain aerial photographs to execute a site inspection of the company's premises under the Clean Air Act. The Environmental Protection Agency did not infringe on either prong of the established 'reasonable expectation of privacy' test when it overflew the company's property without first obtaining a warrant, even though the property was surrounded by a fence. Similarly, in *California v. Ciraolo*[16] the data obtained from a plane hired by the police to fly over a private home without a warrant was permitted. To quote from Ciraolo, the Court found that "[i]n an age where private and commercial flight in the public airway is routine, it is unreasonable for respondent to expect that his marijuana plants were constitutionally protected from being observed by the naked eye"[17]. Three years later, in *Florida v. Riley*[18] the Court affirmed the reasoning that a warrant was not needed to fly a helicopter over a home to collect information.

---

[12] U.S. Constitution (Amendment IV).
[13] 389 U.S. 347 (1967).
[14] Ibid. at 361.
[15] 476 U.S. 227 (1986).
[16] 476 U.S. 207 (1986).
[17] Ibid at 215.
[18] 488 U.S. 445 (1979).

However, law enforcement does not have an unfettered right to use remote sensing technology to collect geo-location information without first obtaining a warrant. For example, in *Kyllo v. U.S*[19] the Supreme Court found that the use of a thermal imaging device aimed at a private home without a warrant violates an individual's right to privacy. Specifically, the Court stated, "[w]here, as here, the Government uses a device that is not in general public use, to explore details of the home that would previously have been unknowable without physical intrusion, the surveillance is a 'search' and is presumptively unreasonable without a warrant"[20].

As can be seen from the examples mentioned above, one of the challenges for the courts is determining what is a reasonable expectation of privacy with respect to data collected from remote sensing technology, particularly as such technology is becoming more common for both commercial and governmental purposes. For example, the Supreme Court has not addressed whether an individual's expectation that law enforcement would not use satellite imagery – as compared to aerial imagery – is unreasonable. However, given the wide availability and use of websites that integrate aerial and satellite imagery, such as Google Earth, it is hard to imagine that such a privacy expectation would be considered reasonable. Similarly, what is considered reasonable will change as new and improved sensors are deployed. For example, one might question whether the Kyllo decision would have been decided differently if thermal imaging devices were as common as they are today[21].

Example: tracking

Another line of location privacy cases concerns the use by law enforcement of location tracking devices. Since the 1983 case *United States v. Knotts*[22], it has been generally accepted that law enforcement can put tracking devices on cars without obtaining a warrant to track an individual while on public roads. In Knotts, the Court found that monitoring a suspect by placing a radio transmitter inside a chemical container that was then transported along public roads, is permitted under the Constitution, as that act did not 'invade any legitimate expectation of privacy'. However, in *United States v. Karo*[23], the Court distinguished the use of a transmitter inside of a private home or building to collect evidence against a suspect from the facts of Knotts. The Court concluded that private residences are places in which people have a strong, reasonable expectation of privacy, which society has accepted. Thus, in Karo the Court found the use of the transmitter without a warrant to be impermissible under the Fourth Amendment.

Improvements in location tracking technology are now forcing U.S. courts to re-examine the legitimacy of the use of tracking devices by law enforcement. For example, law enforcement can now use devices to follow a number of individuals for an extended period of time at a relatively low cost. The question the courts are being asked to address is whether these improvements make the use of the technology a violation of an individual's 'reasonable expectation of privacy'. To put it another way, is it an unreasonable expectation

---

[19] 533 U.S. 27 (2001).

[20] Ibid at 40.

[21] Ibid. at 40 (for further discussion see CCRS (2008). *Report for Congress: Satellite Surveillance: Domestic Issues*, www.fas.org/sgp/crs/intel/RL34421.pdf (accessed on 23/06/2011); I. Kerr and J. McGill (2007). "Emanations, Snoop Dogs and Reasonable Expectation of Privacy". *Criminal Law Quarterly* 52(3), 393-432.

[22] 460 U.S. 276 (1983).

[23] 468 U.S. 705 (1984).

of privacy to expect that law enforcement will not track your movements for an extended period of time, even if on public roads[24]?

There have been a series of cases in the United States on this issue which have resulted in conflicting decisions. In 2007, the United States Court of Appeals for the Seventh Circuit found in *U.S. v. Garcia*[25] that the use of a tracking device on a car while it was in public without a warrant was permitted, concluding that such actions did not constitute a 'search' for Fourth Amendment purposes. Three years later, in *United States v. Pineda-Moreno*[26], the Ninth Circuit reached a similar conclusion.

However, later that year, in *United States v. Maynard*[27], the D.C. Circuit Court of Appeals found that such use of tracking devices was unreasonable. Specifically, the court found that "[s]ociety recognizes Jones' expectation of privacy in his movements over the course of a month as reasonable, and the use of a GPS device to monitor these movements defeated that reasonable expectation"[28]. The Obama administration subsequently challenged the Maynard decision, and the Supreme Court has agreed to hear this case in its fall 2011 session. Hopefully, the Court's decision will provide law enforcement with clearer guidance as how to proceed on this matter.

In addition, a number of bills have been introduced in Congress that would restrict the ability of law enforcement to collect tracking information without first obtaining a warrant. These bills include the Geo-location Privacy and Surveillance Act[29]. It is still too early to know if any of these bills will become law in the United States, but they highlight the increased concern the U.S. Congress has with location privacy.

Stored Communications Act

Law enforcement can also try to track individuals by obtaining records of the location of their location-enabled mobile devices over time. Unfortunately, there is no specific legislation in the United States that directly prescribes what law enforcement needs to show in order to obtain that information from businesses. Courts have been forced to interpret how the Stored Communications Act (SCA) applies to geo-location information. The purpose of the SCA is to address when the government can compel the disclosure of 'a record or other information pertaining to a subscriber to or customer of a service' held by providers of *electronic communications* [emphasis added][30]. It was intended to address issues such as wiretapping or disclosing logs of whom an individual calls. Since geo-location information is not specifically covered under the SCA, the question arises whether law enforcement must show probable cause to obtain a warrant to collect geo-location information – as is

[24] J.C. White (2003). *People, Not Places; A Policy Framework for Analyzing Location Privacy Issues*, http://www.epic.org/privacy/location/jwhitelocationprivacy.pdf (accessed on 23/06/2011).

[25] 474 F.3d 994 (7th Cir. 2007).

[26] 591 F.3d 1212 (9th Cir. 2010).

[27] No. 08-3030 U.S. Court of Appeals, District of Columbia Circuit (August 6, 2010), http://pacer.cadc.uscourts.gov/common/opinions/201008/08-3030-1259298.pdf (accessed on 17/10/2011).

[28] Ibid at 32.

[29] See http://thomas.loc.gov/cgi-bin/bdquery/z?d112:s.1212: (accessed on 17/10/2011).

[30] 18 U.S.C. 2703 (c).

generally required in order to listen in on a call – or whether a lower standard is permitted under the SCA[31].

Courts in the United States have struggled to apply the SCA to geo-location information. Some courts have found that this information is like a pin number and law enforcement needs to obtain a warrant before compelling disclosure of any geo-location data, historical or real-time. Others have found that it is simply a subscriber record and therefore under the SCA the government only needs to show 'specific and articulable facts', which is considered to be a much lower standard[32].

The issue has become critical because of the abundance of geo-location information that is now being collected on individuals by businesses. In September 2010, the Third Circuit Court found that a judge may – but is not required to – ask law enforcement to show probable cause for a warrant before asking a telecommunications company to turn over historical geo-location information on its customers under the SCA. In the *Matter of the Application of the United States of America For An Order Directing A Provider of Electronic Communication Service To Disclose Records to the Government, No. 08-4237*[33], the Obama administration had argued that probable cause was not required under the Act to acquire historical geo-location data – even though 'historical data' could literally be seconds old. Rather, the administration argued all that was required was a showing that 'there are reasonable grounds to believe.... the records or other information was relevant and material to an ongoing investigation' – a much lower standard. Opponents in the case argued that under the SCA a warrant was always required. In finding a judge may ask for probable cause, the court did not agree with either of the arguments.

The Third Circuit Court's inability to make a firm decision on the matter highlights both the ambiguity within the SCA itself with regards to what is required to obtain location-related records, as well as the complexity of the issues involved. Unfortunately, its decision provides little help to lower court judges, law enforcement or the growing number of companies that are collecting this information on a regular basis.

Partly as a result of this uncertainty, in September 2010, Congress began hearings on the SCA and related legislation in an effort to make the law current with existing technology. In May 2011, Senator Patrick Leahy introduced legislation that would require law enforcement to obtain a warrant before compelling a company to disclose its customer's geo-location information[34]. The bill would also broaden the types of companies included under the legislation, to include those that provide remote computing and geo-location information services. However, one wonders by the time such legislation has passed, whether technology will have already made it outdated as well.

---

31 See e.g. Center for Democracy and Technology (2006). *Digital Search and Seizure: Updating Privacy Protections to Keep Pace with Technology*, https://www.cdt.org/publications/digital-search-and-seizure.pdf (accessed on 17/10/2011).

32 See e.g. *People v. Weaver*, 909 V.E. 2d 1995 (N.Y. 2009); *State v. Jackson* 76 P.3d 217 (Wash 2003); *State v. Campbell*, 759 P.2d 1040 (Or. 1988); *Osborn v. State* 44 P.3d 523 (Nev. 2002).

33 3rd Cir 2010.

34 See http://www.govtrack.us/congress/billtext.xpd?bill=s112-1011 (accessed on 17/10/2011).

Businesses and location privacy

Current developments

In the previous sections, the collection and use of geo-location information by government authorities and law enforcement were discussed. However, privacy concerns associated with an individual's location also stem from the activities of the private sector. In the United States, businesses are not subject to the Privacy Act of 1974 or the FOIA. However, they are subject to a variety of regulatory regimes that are intended to protect consumer privacy. Each of these is designed to protect personal information, such as bank accounts, credit scores or health records[35]. Since each law applies to a specific domain, there is no existing legislation that protects a consumer's location or geographic data under a general regime.

Not surprisingly however, given the rapid development of location based services and related technology, privacy issues associated with geo-location information are starting to receive more attention in Congress. In February 2010, the U.S. House of Representatives held hearings specifically on location privacy as part of a broader review of Internet privacy. These hearings resulted in both Representative Rick Boucher[36] and Representative Bobby Rush[37] drafting legislation to protect consumer personal data. Both of these bills included 'precise geo-location information' as sensitive information that would be subject to greater protection with regards to collection, use and distribution, although neither defined 'precise geo-location information'.

Representative Rush's bill did not progress far and Representative Boucher's bill was not formally introduced. However, in 2011, a number of new bills were introduced that aim to regulate the collection, use or transfer of an individual's geo-location for privacy purposes. These bills can be broken down into two types. One set of bills attempts to regulate geo-location as part of a larger effort to protect internet privacy. Under these bills, companies that collect information over the internet would be required to follow the same procedures in protecting an individual's precise geo-location information as they would with respect to an individual's social security number or unique identifier associated with a computer. These bills require specific notice provisions prior to the collection of precise geo-location information that would be used in a way to identify an individual. In addition, the individual should have access to any information that is collected as well as the ability to control the transfer of information to third parties. One example of this type of bill would be the Commercial Privacy Bill of Rights Act of 2011[38] introduced by Senator John Kerry[39] and Senator John McCain[40].

The second set of bills restricts the possibilities for tracking an individual's location and focuses more on mobile devices. Each bill would require the consent of an individual before geo-location information is collected, received, disclosed, recorded or obtained

[35] Health Insurance Portability and Accountability Act (health records); Gramm-Leach-Bliley Act (financial records); Fair Credit Reporting Act (credit information); Children's Online Privacy Protection Act (information on minors collected over the Internet); Video Privacy Protection Act.

[36] Representing the democrats for the State of Virginia.

[37] Representing the democrats for the State of Illinois.

[38] See http://thomas.loc.gov/cgi-bin/bdquery/z?d112:s.00799: (accessed on 17/10/2011).

[39] Representing the democrats for the State of Massachusetts.

[40] Representing the republicans for the State of Arizona.

unless otherwise required by law or for public safety. These bills include Senator Franken's[41] Location Privacy Protection Act of 2011[42].

Even in the absence of legislation, the federal government probably already has the ability to regulate what companies can do with consumers' geo-location information. Specifically, Section 5 of the Federal Trade Commission Act grants the Federal Trade Commission (FTC) broad enforcement authority to protect consumers from unfair trade practices[43]. Initially, the FTC used this enforcement authority with respect to privacy to ensure that a company complied with the stated privacy statements on its web site on how it protects its customers' sensitive personal data, such as a credit card information or social security numbers[44]. The FTC then began to take enforcement actions against companies that lost (or had stolen) their customers' sensitive personal data, even if such loss did not violate a stated privacy policy with respect to such data. Enforcement actions included both fines and other civil penalties. Specifically, the FTC found that the failure to adequately protect sensitive personal data was an unfair trade practice, irrespective of whether a company had a privacy policy for such data[45].

Most recently, the FTC appears to have expanded its enforcement actions to protect customer data beyond sensitive information as social security numbers, credit card information or credit history. For example, in a recent enforcement action against Twitter, the FTC found that Twitter failed to provide 'reasonable safeguards to protect user information from unauthorized users'[46]. This information included passwords to send messages ('tweets'), but not information that had previously been considered sensitive. Apparently, the FTC is trying to send a message to businesses that even non-sensitive data associated with an individual must be protected if it is stored in a non-public place.

Discussion

There are many unanswered questions with respect to the proposed legislation on the protection of geo-location information. For example, in many bills the term 'geospatial information' is not defined. Similarly, the bills fail to take into account that geo-location information is different from other types of information considered sensitive, such as social security numbers or sexual orientation. These differences include:

- Defining 'geo-location information' is much more difficult than defining 'social security number' or 'religious affiliation'. For example, is an individual's privacy violated if his or her location is disclosed to within a few feet, or a certain region, such as a city or county? This has become even more important after a California court recently held that a zip code is 'personally identifiable information' under California law[47].

[41] Representing the democrats for the State of Minnesota.
[42] See http://thomas.loc.gov/cgi-bin/query/z?c112:S.1223: (accessed on 17/10/2011).
[43] See http://www.ftc.gov/opp/gpra/append1.shtm (accessed on 17/10/2011).
[44] See http://www.ftc.gov/opa/2008/03/studlend.shtm (accessed on 17/10/2011).
[45] See http://www.ftc.gov/opa/2010/03/davebusters.shtm (accessed on 17/10/2011).
[46] See http://www.ftc.gov/os/caselist/0923093/100624twitteranal.pdf (accessed on 17/10/2011).
[47] Pineda v. Williams-Sonoma Stores 51 Cal.4th 524 (2011).

- Geo-location data can and will increasingly be used to provide a broad range of important services, such as the use by emergency responders in disaster relief and by the medical community to predict patterns of infectious diseases. In many such instances it will be nearly impossible to provide prior, timely and accurate notice to a consumer as to how such information will be used, yet this would be required under certain proposed legislation.
- An individual's location is collected, geocoded, electronically stored and transmitted many times a day, through such devices as red light camera, security closed circuit television (CCTV), automobile smart tags and office building key cards. Trying to draft laws and regulations that regulate the various methods used to collect and use an individual's precise geo-location information raises numerous operational challenges that have yet to be fully considered. For example, some legislation would require obtaining prior consent from an individual before collecting his or her geo-location information. However, such consent would be impractical with respect to CCTV or red light cameras.
- As described above, the U.S. Supreme Court has held on a number of occasions that there is no 'reasonable expectation of privacy' in a public place. These decisions have primarily been related to Fourth Amendment 'search and seizure' cases; however they are generally accepted in a variety of contexts. Efforts by Congress to deem an individual's 'precise geo-location information' while in a public place as sensitive could have a profound impact on this analysis, particularly with respect to law enforcement and national security. For example, criminal defence lawyers might argue that Congress has created a 'reasonable expectation of privacy' in a public place, and therefore law enforcement cannot use a vehicle to follow a suspect without first obtaining a warrant.

With regard to the activities of the FTC, it is still too early to determine whether it will take enforcement action with respect to the security of geo-location information without legislation granting it specific authority to do so. However, given the recent media attention in the United States on privacy concerns[48] associated with geo-location information and high-profile congressional hearings on the subject, it is fair to assume that the FTC will consider doing so in appropriate cases. In fact, continued reports on how easily hackers can obtain geo-location information from smart phones and other mobile devices will most likely increase the probability of scrutiny[49].

The FTC's enforcement actions regarding privacy protection have primarily focused on the failure of companies to adequately protect personal data from improper use or access once it has been collected. Through its actions and other pronouncements, the FTC has stated clearly that the failure of companies to develop a comprehensive security plan with respect to sensitive data is an unfair trading practice. This concept of a security plan is not new; it is similar to that required under the Health Insurance Portability and Accountability Act[50] and the Gramm-Leach-Bliley Act[51]. According to the FTC, a security plan should be comprehensive and in writing and should be based upon the size of the company and the

---

[48] See e.g. http://online.wsj.com/public/page/what-they-know-digital-privacy.html (accessed on 17/10/2011).
[49] See http://articles.latimes.com/2010/jul/04/business/la-fi-mobile-security-20100704 (accessed on 17/10/2011).
[50] See http://aspe.hhs.gov/admnsimp/pl104191.htm (accessed on 17/10/2011).
[51] See http://www.gpo.gov/fdsys/pkg/PLAW-106publ102/pdf/PLAW-106publ102.pdf (accessed on 17/10/2011).

sensitivity of the data. The FTC does not expect a security plan to prepare for all contingencies. However, it should include provisions on the following:

- Risk assessments: The objective should be to identify risks and determine how to mitigate those risks. The FTC has made it clear that not every risk has to be identified; only those that are reasonably foreseeable.
- Employee training on information security issues: Employees have frequently proven to be responsible for lost or stolen personal data. A plan should provide that access to data is limited to those employees with a legitimate need, and those with access should be trained on proper security procedures.
- Disposal procedures for data: Data should not be retained longer than necessary.
- Regular monitoring and updating of the security plan: Security threats and technology change over time. According to the FTC, security plans should be updated periodically in anticipation of both.
- Plans to report, respond to and manage security incidents: Any data security plan should address what happens if data is lost, stolen or misused. The plan should detail who should be alerted and what steps should be taken to mitigate further damage.
- Policies in hiring third parties: There have been a number of instances where third party contractors have lost or stolen personal data. A comprehensive security program should address due diligence concerning vendors and contractors and also should include required data security provisions in contracts.
- Designated employee(s) responsible for data security: The FTC has stated that designating a senior level employee as the person responsible for implementing and maintaining a plan is the best way to make sure that this plan is followed.
- Information security practices: Information security should be in keeping with best practices, such as making administrative passwords difficult to guess, suspending or disabling passwords after a reasonable number of login attempts and enforcing periodic changes to passwords. Even though there is not yet a legal requirement to do so, companies that collect or store geo-location information on individuals in the United States should consider implementing a similar security plan to protect that information from unauthorized disclosures. First, if legislation is enacted that empowers the FTC to deal with geo-location privacy, such a plan is likely to become required[52]. Moreover, it is increasingly a good business practice, particularly given the media's attention for privacy and data breaches.

Conclusion

The United States is inching closer to developing a comprehensive legal and policy framework that addresses location privacy. However, as discussed above, there are a number of factors that still must be considered in developing a privacy regime for geo-location information. Therefore, any such regime should recognize a number of considerations.

---

[52] See http://www.ftc.gov/opa/2010/12/privacyreport.shtm (accessed on 17/10/2011).

First, the concept of location privacy is a relatively new one. One's sense of location privacy will depend in part upon one's age, gender and the culture one was raised in. As a result, there will be no easy answer to address all the concerns that are expressed.

Also, geo-location information is collected every day by a variety of sensors. Therefore any attempt to regulate this information must be narrowly tailored so as not to impinge on the collection of data that is being collected and used for purposes that do not pose a realistic privacy risk. The challenge will be in writing laws and regulations that define which businesses and uses are covered and which should be considered exempt.

In addition, the technology is just now developing. If the technology is over-regulated before it matures, many important uses may not develop. While there are certainly privacy risks associated with geo-location information, there are also a number of benefits. Before regulating location privacy, an assessment should be made as to what the opportunity cost would be of not having the data available. In addition to the obvious benefits of some geo-location information for society (e.g. in disaster management of for the deployment of emergency services), one should also not ignore its economic potential.

Geospatial technology has the potential to become a key component for economic growth. Given the expected growth in such technology as smartphones and mobile tablets, the geospatial ecosystem required to support corresponding growth in mobile applications will create numerous jobs and increase tax revenues. Premature regulation of geospatial technology without a better understanding of the true nature of location privacy would limit these growth opportunities.

At the same time, it is important to assess the impact of new legislation on location privacy on the existing practices of businesses that use geo-location information. Businesses have been collecting customer geo-location information, such as addresses or zip codes, for years. As a result, this type of data is likely being stored in numerous data bases throughout a company. Trying to harmonise current and future business practices so as to protect this information will prove difficult for many companies and impossible for some.

Finally, when designing legislation on location privacy, Congress should also keep in mind that there may be other means to address location privacy concerns. For example, one of the main concerns is the use of geo-location information to stalk an individual. Such concerns may be better addressed through amendments to the criminal code rather than including loosely-defined 'geo-location information' in a broad privacy regime.

The Processing of Location Data under the European Data Protection Legal Framework: Behavioural Profiling via Location Tracking

Eleni Kosta[1] *and Mark N. Gasson*[2]

Introduction

The miniaturisation of GPS devices and their inevitable inclusion in mobile handsets has generated a new era of information disclosure, and new services which encourage people to reveal where they are at any time in the name of safety, convenience or for social use are becoming widely adopted. Services that are based on the location information of users cover for example services that provide the users with traffic information or offer guidance to drivers, direct marketing services based on the location data of users, tracking services for children or for people with Dementia or Alzheimer. These are commonly known as location based services.

The term location based services is very broad and does not only cover the value added services that are based on the processing of location data other than traffic data offered via a public communications network or in a publicly available electronic communications service, within the scope of the ePrivacy Directive[3], as will be elaborated in this chapter. It covers different types of services that are delivered using various wireless technologies and are based on the processing of location information, falling both within and beyond the definition of location data of the ePrivacy Directive. Location based services can for instance be offered via the following technologies: satellite-based positioning systems such as the Global Positioning System (GPS); sensor-based systems such as face recognition systems and licence-plate scanners for vehicles; wireless technologies, such as Radio Frequency Identification (RFID), WiFi or Bluetooth; cell-based mobile communication networks such as GSM and UMTS; and chip-card-based systems, such as credit cards[4].

Thus, the term location based services is rather broad. However, in the course of this chapter, the term location based services will be used as referring to the value added services that are offered based on the processing of location data of users and are governed by Article 9 of the ePrivacy Directive.

New generations of mobile handsets are ideal for the deployment of location based services because not only do they have the technological capability to capture and send data, but they are generally carried by an individual everywhere they go, practically at all times. However, rather than being limited to providing information based on isolated and

---

[1] Interdisciplinary Centre for Law and ICT, KU Leuven, eleni.kosta@law.kuleuven.be.

[2] School of Systems Engineering, University of Reading, m.n.gasson@reading.ac.uk.

[3] European Parliament and Council (2002). Directive 2002/58/EC of 12 July 2002 concerning the processing of personal data and the protection of privacy in the electronic communications sector (Directive on privacy and electronic communications), *OJ L* 201, 37.

[4] C. Cuijpers and B.J. Koops (2008). "How fragmentation in European law undermines consumer protection: the case of location-based services". *European Law Review* 33, 880-897.

discrete geographic locations, such GPS enabled devices can provide advanced services that greatly customise information to the user by, for example, monitoring where the user goes at any given time. Data mining can be applied to this wealth of data to find what routines/ habits they have, where they tend to go and other user-specific pieces of information. From this, profiles can subsequently be drawn. In the case of monitoring multiple users, social and business networks can also be inferred. The advantage to the user is that, for example, when travelling to a new city, the user can ask the service individual specific, yet abstract questions: '*where are restaurants that I would like?*' or '*where do people like me go to here?*' and so on.

While these services will undoubtedly be useful, the seemingly harmless data may not just reveal such things as where you are or have been; persistent collection of data can expose aspects of one's private life that at first glance an individual may not realise. It is not just where individuals live, where they work or when they go for a coffee; it is possible to aggregate all these pieces of information and use data mining techniques to extract a 'behavioural profile' from the data. Problems could arise if for example this information is used by third parties to vary their services or prices specific to the individual, in some cases to their detriment. This is especially so given that the end 'user' may have no idea that this is happening, or indeed how it is happening.

To further explore the information that can be revealed about a user from the collection and processing of his location data, and to assess the potential impact on the privacy of the user in light of the European data protection legislation, four people from three different European Member States took part in an experiment during the month of April 2009. Over this period they were persistently tracked via GPS enabled mobile phones and their location data were stored in a central database for automated and manual processing. The aim of this processing was a first attempt to mine new information from the data relevant to forming behavioural profiling of the individuals. This was based on where they had been and was used to investigate in what ways the data collected over this experimental period supports our hypotheses that it is an adequate amount of data with which to draw a simple profile of both individual and group behaviour and that the data can be highly privacy invasive.

In this chapter, following a description of the experiment, the current European legal framework on the processing of location data is examined and the implications it has on the deployment of location based services are discussed.

Data collection and processing

The design of the overall system employed in the experiment was the result of multiple iterations based on conflicting factors such as usability, cost, availability and functionality. The involved process is not detailed here – instead we briefly describe the resulting implementation.

Each user was issued with a HTC 6500 handset with integrated GPS, fingerprint reader and full size SD slots for additional peripheral devices. Running Windows Mobile 6.1, the unit provided an ideal development platform for this study, simplifying the hardware solution into a mainly integrated device which was less prone to user damage. Two pieces of custom software were run on all devices. The first logged each time the users authenticated themselves with the device (typically through the fingerprint reader). The second polled the internal GPS unit at 10 second intervals for updated data. The software then calculated the distance travelled and if it was greater than 100m, the cell tower details from the Windows Mobile Radio Interface layer were collected along with the last time of authentication from the log. Since the cell tower information only gives details about the id of the cell tower, the opencellid.org service was used to retrieve the GPS co-ordinates of the cell tower (if available). All of these discrete events were then sent for storage in a MySQL database as detailed in the following.

Table 1: Example temporal and geographic data from a handset as stored in a MySQL database

Field	Type	Example Data
Row	bigint(20)	32974
Latitude	double	50.86544374
Longitude	double	4.65134145
TowerId	int(11)	41766
MobileCountryCode	int(11)	206
MobileNetworkCode	int(11)	20
LocationAreaCode	int(11)	303
cellLatitude	double	50.872253
cellLongitude	double	4.660578
IDTime	varchar(14)	1241033658
ID	varchar(11)	'unique ID'
Time	varchar(14)	1241035074
Realtime	varchar(40)	Wed, Apr 29, 2009, 19:57:54 GMT
POI	tinyint(1)	0
Use	tinyint(1)	1

In Table 1, 'Row' is simply the incremental row identifier, 'IDTime' is the timestamp of the last authentication, and 'ID' is the unique user identifier. 'Time' refers to the time the data was sent in Unix timestamp form and 'Realtime' is a human readable version of that. 'Use' specifies whether a data point should be used in the data processing or not, and is designed

Figure 1: A plot of the actual points a location has been logged (each indicated with a blue flag). Note the 'Toggle Status' link in the information panel

for removal of spurious data points (which can occur if the GPS signal is very poor). The front end to the database is a bespoke PHP enabled viewer which allows user definable algorithms to be run on the data, the results of which are then displayed using the Google Map Application Program Interface (API). This allows simple functions such as displaying the last known location of every user, or the movement activity of any user on any day. Equally the actual location points can be plotted, with an option to toggle their 'use' indicator on and off to remove spurious data points if need be (Figure 1).

User profiling from GPS data

It should be noted that the purpose of this study is not to implement a complex and fully automated system that comprehensively profiles the participants. Instead the aim is to examine the types of privacy invasive information we can draw from the location data of the users that are collected over a relatively short period, and so investigate in what ways such systems could be privacy invasive if the user of a location based service agreed to persistent tracking. As such, the interactive interface was used during the processing phase to both run various automated tasks and to enable manual inspection of the data.

One of the key tools for initial mining was the generation of 'Points of Interest' (PoI) from the data (see Figure 2). This used a technique of calculating the time between subsequent location points and concluding that a PoI exists if the elapsed time is over a preset threshold value. In this study a threshold of 3 minutes was deemed adequate[5].

Figure 2: Indicators marking automatically generated 'Points of Interest' with optional satellite overlay on the map

[5] For this study the techniques specified by Ashbrook and Starner were used: D. Ashbrook and T. Starner (2003). "Using GPS to learn significant locations and predict movement across multiple users". *Personal and Ubiquitous Computing* 7(5), 275-286.

The next phase of the process was to determine the location identified as being of interest. It is expected that in a fully fledged profiling system this would be an automated process utilising multiple online Internet resources such as maps, databases and business listings. Although some algorithms have been utilised to help mine the data in this study, much qualitative assessment, manual processing, and in some cases survey of the users has also been performed in order to aid in the exploitation and to expedite the process. In the case of the data shown in Figure 2, from an online map resource and the times the points occurred we can infer that the user arrives at work (Systems Engineering Building at the University of Reading, UK) at around 09:00, takes lunch in the university cafeteria from 11:45 until 12:20,

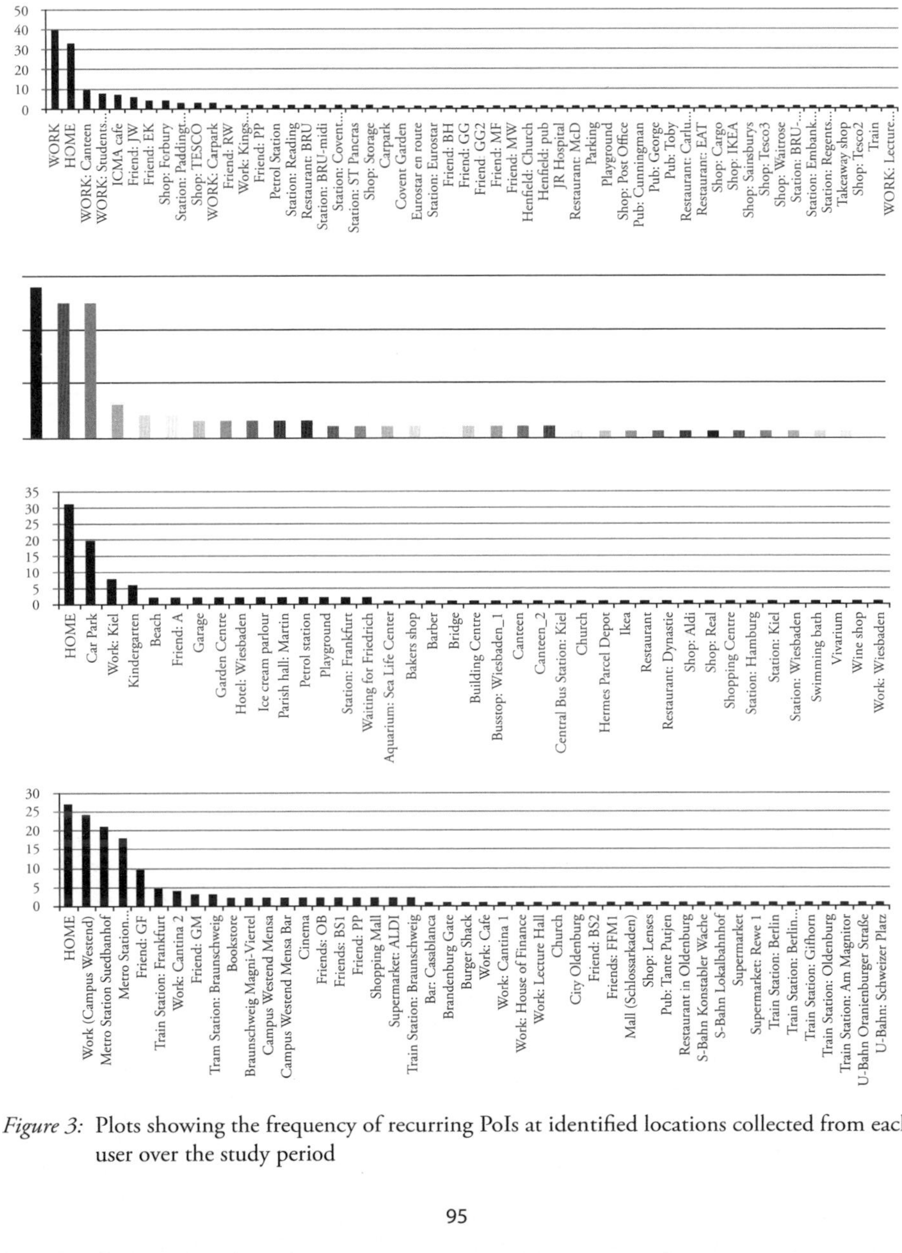

Figure 3: Plots showing the frequency of recurring PoIs at identified locations collected from each user over the study period

then presumably takes a drink at the cafe in the Business centre before returning to work. This person then leaves work at 21:45.

This study is based on the collection of GPS/location data of the four participants. By calculating the frequency of recurring PoIs, which are based on the location information of the users, the key locations for each user become apparent. These are given in Figure 3.

If we consider how PoIs are generated, it is evident that key locations are those frequently visited and so will attract a 'cluster' of PoI points (Figure 4). In some cases, leaving and returning from a work building several times during the day may be commonplace – such as leaving and returning from lunch – and so will register multiple times per (work) day. This may well be in contrast to a residence which may only be left and returned to once a day.

Figure 4: Clusters of PoIs at key work locations collected from one user over the study period

The use of temporal data can also serve to disclose the identity of some PoIs – for example the location occurring most often first and last thing during the day is likely to be the residence. Further analysis can also shed light on typical work day routine, and give an indication of job type undertaken by the users.

The graphs in Figure 5 show the average daily routine for three of the four users, with error bars showing the standard deviation from these averages. Notably none of these three users have a rigid routine, demonstrated by the notable deviations from the average. They do however all keep core working hours, and in some cases this involves work over the weekend. Further processing of the location data shows two of the users were engaged in foreign travel, and both appeared to do so mostly in a work context. Notably User 4 has a dependency on public transport while the others evidently own their own car which they predominantly use.

Whilst care needs to be taken in interpreting these results, especially due to influencing factors such as the ability to work from home – given that the location can be determined, a likely job profile could be drawn for these users. Notably, User 3 lacks a definable daily routine, and visits an office environment only a few times during the study. This user does not conform to the expected profile of an office worker, however the lack of routine also

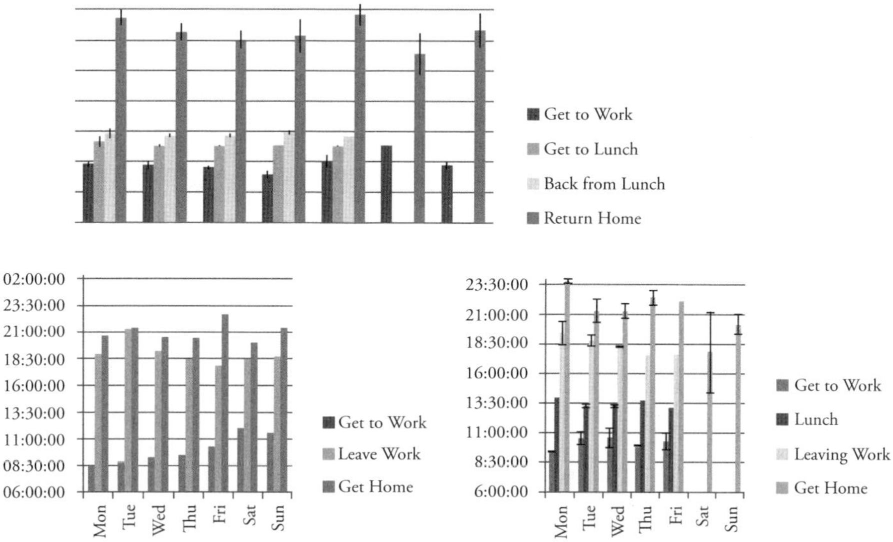

Figure 5: Plots showing the average times of key events during the week for User 1, 2 and 4 respectively. Error bars show one standard deviation from these average values

does not indicate that of a housewife/husband. Given the data, it is not possible to clarify the exact work status of this user.

Data protection and location data

As all the participants in the experiment are European citizens and the collection and processing of the data takes place in Europe, it offers an excellent opportunity to examine the current European legal framework on the processing of location data.

Location data

The ePrivacy Directive defines location data as

> "any data processed in an electronic communications network or by an electronic communications service, indicating the geographic position of the terminal equipment of a user of a publicly available electronic communications service"[6].

The definition was recently amended by a 2009 Directive in order to cover not only data that are processed in an electronic communications network, but also data processed by

[6] Article 2(c) ePrivacy Directive.

an electronic communications service[7]. Location data may, thus, refer to the geographic coordinates of the terminal equipment of a user, i.e. its latitude, longitude and altitude, to the identification of the network cell in which the terminal equipment is located at a given time, to the level of accuracy of the information that relates to the location of the user, as well as to the time this information is recorded[8]. The exploitation of PoIs is based on precisely such location data of the users.

With regard to the nature of location data as personal data, the Article 29 Working Party[9] has taken the position that location data always relate to an identified or identifiable natural person and are therefore always personal data[10]. This statement is correct insofar as location data are treated in the context of the ePrivacy Directive and are examined in compliance to the definition of Article 2(c) of the Directive, which confines location data to data 'indicating the geographic position of the terminal equipment of a user of a publicly available electronic communications service'[11]. Therefore location information relating to objects that cannot be linked directly or indirectly to a natural person does not qualify as location data under the meaning of Article 2(c). Cuijpers and Koops argue on this issue that 'it is questionable whether this statement [of the Article 29 Working Party] is correct, since location data can also relate to objects that are not linkable to individual natural persons'[12]. As an example, they mention location information that refers to places and machines in a fully automated warehouse that is used only by machines[13]. This statement is correctly taken out of the scope of the ePrivacy Directive and could refer in general to location information that can relate not only to natural persons, but also to objects. However, within the limited scope of the ePrivacy Directive and in accordance to the restrictive definition of location data in its Article 2(c), location data will always be personal data, as they indicate the geographic position of the terminal equipment of a user of a publicly available electronic communications service.

Besides the location data that qualify also as traffic data and have to be processed in accordance with Article 6 of the ePrivacy Directive, there are location data that are not traffic data and are 'not processed for the purpose of the conveyance of a communication on an electronic communications network or for the billing thereof'[14]. These data are referred to in the ePrivacy Directive as location data other than traffic data and are processed for the

[7] Council of the European Union (2009). Common Position (EC) No 16/2009 adopted by the Council on 16 February 2009 with a view to the adoption of a Directive 2009/.../EC of the European Parliament and of the Council of ... amending Directive 2002/22/EC on universal service and users' rights relating to electronic communications networks, Directive 2002/58/EC concerning the processing of personal data and the protection of privacy in the electronic communications sector and Regulation (EC) No 2006/2004 on consumer protection cooperation protection laws, *OJ C* 103 E/40.

[8] Recital 14 ePrivacy Directive.

[9] The Article 29 Working Party was established by the 1995 Data Protection Directive (European Parliament and Council (1995). Directive of 24 October 1995 on the protection of individuals with regard to the processing of personal data and on the free movement of such data, *OJ L* 281, 31). Its tasks include promoting the uniform application of the general principles of the privacy directives, and advising the European Commission on matters relating to the protection of persons with regard to the processing of personal data. (see http://ec.europa.eu/justice/policies/privacy/workinggroup/index_en.htm).

[10] Article 29 Working Party (2005). *Opinion 5/2005 on the use of location data with a view to providing value added services*, http://ec.europa.eu/justice/policies/privacy/docs/wpdocs/2005/wp115_en.pdf (accessed on 19/11/2011).

[11] Article 2(c) ePrivacy Directive.

[12] C. Cuijpers and B.J. Koops (2008). "How fragmentation in European law undermines consumer protection: the case of location-based services." *European Law Review* 33, 880-897.

[13] Ibid.

[14] Article 2(b) ePrivacy Directive.

provision of value added services[15] that are based on the location of the user. In the study presented here, the location data collected via GPS would qualify as location data rather than traffic data. Such value added services are commonly known as location based services and cover for instance services that provide the users with traffic information or offer guidance to drivers[16], direct marketing services based on the location data of users, tracking services for children or for people with Dementia or Alzheimer.

In the context of this study, we wished to examine the types of information that can be revealed about the user via the exploitation of PoIs (and subsequently the location of the user) and draw some conclusions on the ease to build user profiles. The processing of PoIs revealed for instance information about the residence of the user, as this can be easily concluded based on the location information combined with the time when the user leaves and returns to a certain location.

Inferring relationships

The processing of location data of users can reveal rich information on their relations with other people, especially if these people are also subscribers to location based services and are being profiled. In the case of this study, by looking for instances when two users were in close proximity of each other, a social or business relationship could be inferred. Figure 6 shows the route of two of the users – they use different tube stations, but then arrive at the same location before walking together to a second location and back again (note the time information is not shown). By observing the times at which this happened, and the end locations (PoIs), especially if it happens multiple times, we can infer whether this is a social

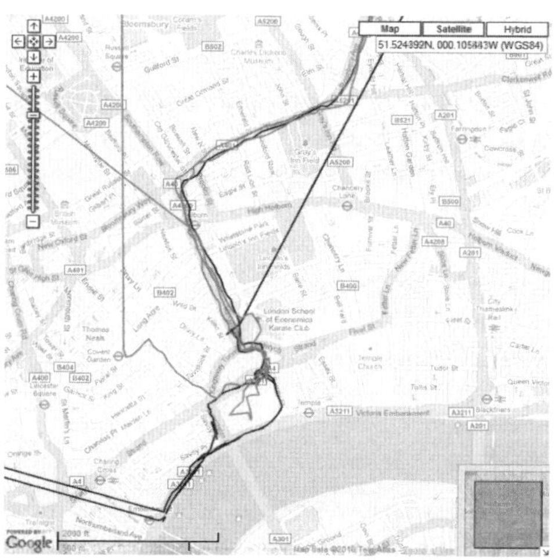

Figure 6: Inferring a relationship by observing occasions of spatial and temporal unity

[15] Value added services are defined in Article 2(g) of the ePrivacy Directive as 'any service which requires the processing of traffic data or location data other than traffic data beyond what is necessary for the transmission of a communication or the billing thereof'.

[16] Recital 35 ePrivacy Directive.

or business relationship. In this case the users meet from 11:02 to 16:24 at King's College in London, UK. This would imply a business relationship, and would also indicate that there should be a degree of compatibility between the working lives of both users.

Sensitive data

Sensitive data are defined in Article 8(1) of the Data Protection Directive[17] and cover 'personal data revealing racial or ethnic origin, political opinions, religious or philosophical beliefs, trade-union membership, and the processing of data concerning health or sex life'.

The processing of sensitive data is in principle prohibited and is only allowed under specific conditions and strict safeguards. Any information about a living individual that includes facts and intentions or opinions about any of these categories is considered sensitive data. Given that the processing of sensitive data is in principle prohibited and is only allowed under specific conditions and strict safeguards, it is interesting to examine whether location data can potentially reveal information about any of these aspects.

Location specific indicators and social interactions are a potential source of information regarding inferring someone's sexual preferences. In the case of the data in this study, no such obvious indicators exist. Moreover, simply attending a gay-bar for example is not sufficient to draw conclusions – rather it would have to be a wider picture drawn from disparate elements. Notably drawing the conclusion that a person is heterosexual is as relevant as inferring homosexuality or bisexuality, but is in itself potentially still non-trivial.

Routinely attending a doctor's office, hospital or specific health clinic is potentially an indicator of a health issue – although not necessarily that of the person being tracked as they may be accompanying someone else. Conversely, not attending a health professional, such as a dentist or optician, on a routine basis could be an indicator of general apathy regarding health.

Movement patterns may also be an indicator of underlying medical problems. More concretely, the amount of physical exercise a person takes has been directly linked to risk factors associated with ailments such as heart disease. Exercise can be quite apparent in location data through walking or running – although attending a gym is another, albeit less measurable, sign.

The chief medical officer in the UK recommends that adults should do a minimum of 30 minutes moderate-intensity physical activity, five days a week. On this basis, it would seem that none of the four users in this study take enough exercise, and so their health could be at risk. This is of course making the assumption that they do not e.g. exercise at home, or have access at work to a gym that would not be apparent in the location data.

As location data can reveal information for instance on the health or sex life of the user, it should be treated as sensitive data. However, the fact that the processing of location data is in principle prohibited and is only allowed under strict conditions specified in Article 8 of the Data Protection Directive, raises questions on the practical difficulties in the deployment of any kind of location based services. Location data reveal context-rich information about the user that can easily be considered as sensitive data. However, prohibiting their processing,

[17] European Parliament and Council (1995), Directive 95/46/EC of the European Parliament and of the Council of 24 October 1995 on the protection of individuals with regard to the processing of personal data and on the free movement of such data, *OJ L* 281, 31.

as it would be dictated by Article 8 of the Data Protection Directive, would actually render the deployment of location based services impossible.

Processing of location data for the provision of location based services

The ePrivacy Directive in its Article 9 regulates only a fraction of location based services, namely the ones that rely on the processing of location data other than traffic data offered via a public communications network or in a publicly available electronic communications service[18]. Clearly, location based services that are offered to the members of a private network are not governed by the provisions of Article 9 of the ePrivacy Directive. For example, Article 9 does not cover location data that are transmitted via enterprise networks aimed for a private user group, or data collected and transmitted via infrared signals or GPS signals in combination with a private secured wireless LAN[19]. However, it is not always easy to judge whether the processing of location data for the provision of a location based service falls under the protective ambit of Article 9 of the ePrivacy Directive.

Currently, as expressed also during the adoption of the 2009 Directive amending the privacy and electronic communications framework, the legislative bodies of the European Union wish to uphold the criterion that the European electronic communications regulatory framework applies only to *publicly* available electronic communications services and *public* communications networks[20]. Several of the enabling technologies for location based services, such as RFID, WiFi, Bluetooth or sensor-based technologies, belong to the grey zone between private and public networks and services. Therefore, the decision on the applicability of the ePrivacy Directive on them is not clear. Cuijpers and Koops have suggested as a possible solution the abolishment of the requirement that a service and a network need to be 'public' for the ePrivacy Directive. This means that in the future the ePrivacy Directive could be amended in order to cover not only public but also private networks and services[21]. Cuijpers and Pekárek see two alternative approaches that could clarify the applicability issues of the ePrivacy Directive to location based services: either to extend the ePrivacy Directive by amending its scope, or to restrict the scope of the ePrivacy Directive and extend the scope of the Data Protection Directive, so that the latter

---

[18] Article 9(1) ePrivacy Directive.

[19] R. Steidle (2009). "Datenschutz bei Nutzung von Location Based Services im Unternehmen." *Multimedia und Recht*, 167-170.

[20] The Data Retention Directive also applies to 'providers of publicly available electronic communications services or public communications networks'. In the context of the discussion on the scope of application of the European regulatory framework on electronic communications, the European Commission clarified that only public electronic communications networks and services are covered by it. A. Klabunde, the Head of Sector DG INFSO, referred to the issue stating that: "According to the scope of the eCommunications framework, this definition [i.e. the definition of providers of publicly available electronic communications service or of communications networks] covers economic actors in electronic communications, providers of fixed telephony, mobile telephony, Internet access, Internet e- mail and Internet telephony. Outside the scope are providers of Information society services, such as content services, web services (explicitly excluded from the Framework Directive) and in house communications". (European Commission (2007). *Meeting on Implementation Issues surrounding the Data Retention Directive (Directive 2006/24/EC). Meeting report*, http://www.cullen-international.com/cullen/multi/doc/security/meeting140307_data_retention.pdf (accessed on 11/10/2011).

[21] C. Cuijpers and B.J. Koops (2008). "How fragmentation in European law undermines consumer protection: the case of location-based services." *European Law Review* 33, 880-897.

would govern the processing of location data in all types of location based services[22]. The adoption of the latter approach presents the additional advantage that the processing of location data would not necessarily rely on the consent of the user or the subscriber, the validity of which is questionable in the current deployment of location based applications where the users have to accept the processing of their location data under the conditions offered by the provider in order to be able to enjoy the service.

The limited scope of Article 9 requires that the providers of location based services have to analyse the types of data they are processing, the type of service they are offering and the type of network they are using, in order to find out whether they have to comply with the strict requirements of Article 9; these requirements will be examined below.

Processing of location data other than traffic data

Within the protective ambit of Article 9 of the ePrivacy Directive, location data other than traffic data relating to users or subscribers of public communications networks or publicly available electronic communications services that are used for the provision of a location based service may be processed only when they are made *anonymous*, or when the users or subscribers have given their *consent* to the provision of such a location based service[23].

In any case, the location data may only be used to the extent and for the duration necessary for the provision of the value added service[24]. The users and the subscribers should be given the opportunity to withdraw their consent for the processing of the location data at any time[25] and the withdrawal will be valid for the future. The users and the subscribers should also be offered the choice to temporarily refuse the processing of their location data for each connection to the network or for each transmission of a communication, using simple means and free of charge[26]. This can for instance be done via a switch on/off option of the terminal equipment. The processing of location data may be undertaken only by persons acting under the authority of the network operator, the service provider or the third party providing the value added service and only for the purposes of providing the value added service[27]. It should be pointed out that while the obligations relating to the processing of location data for the provision of value added services cover only the mobile operators (or the electronic communications operators in general), the third parties that are involved with the provision of the value added service on the basis of location data that they receive from the mobile operators are bound by the obligations arising from the Data Protection Directive[28].

[22] C. Cuijpers and M. Pekárek (2011). "The regulation of Location Based Services: Challenges to the European Union Data Protection Regime." *Journal of Location Based Services* 5(4), 223-241.

[23] Article 9(1) ePrivacy Directive. A high-level description of the European framework on location based services can be found at J. Hladjk (2009). "Location Based Services: European data protection rules for mobile commerce". *Privacy and Security Law Report* 8, 21-26.

[24] Article 9(1) ePrivacy Directive.

[25] Ibid.

[26] Article 9(2) ePrivacy Directive.

[27] Article 9(3) ePrivacy Directive.

[28] Article 29 Working Party (2005). *Opinion 5/2005 on the use of location data with a view to providing value added services*, http://ec.europa.eu/justice/policies/privacy/docs/wpdocs/2005/wp115_en.pdf (accessed on 19/11/2011).

Although location data other than traffic data can be processed only for the duration necessary for the provision of the value added service, they can exceptionally be retained for a limited period based on a specific legislative measure taken by the Member States. Such retention is only allowed when it

> "constitutes a necessary, appropriate and proportionate measure within a democratic society to safeguard national security (i.e. State security), defence, public security, and the prevention, investigation, detection and prosecution of criminal offences or of unauthorized use the electronic communications."[29]

In derogation from Article 9, retention of location data is also allowed for a period of six months to two years for law enforcement purposes in accordance with the Data Retention Directive[30].

Provision of information

The ePrivacy Directive explicitly requires that the service provider must, before obtaining the consent, provide the individual with specific information regarding the type of location data that will be processed, regarding the purposes and the duration of the processing, and whether the data will be transmitted to a third party for the purpose of providing the location based service[31]. To the extent that location data are also personal data, the general information obligations that are foreseen in the Data Protection Directive are also applicable. The Article 29 Working Party compiled a comprehensive list on the information that should be provided to the individuals before obtaining their consent for the processing of their location data for the provision of a location based service:

- The identity of the controller and of his representative, if any;
- The purposes of processing;
- The type of location data processed;
- The duration of processing;
- Whether the data will be transmitted to a third party for the purpose of providing the value added service; the right of access to and the right to rectify the data;
- The right of users to withdraw their consent at any time or temporarily refuse the processing of such data, and the conditions on which this right may be exercised;
- The right to cancel the data[32].

[29] Article 15(1) ePrivacy Directive.

[30] European Parliament and Council (2006). Directive 2006/24/EC of 15 March 2006 on the retention of data generated or processed in connection with the provision of publicly available electronic communications services or of public communications networks and amending Directive 2002/58/EC ('Data Retention Directive'), *OJ L* 105, 54.

[31] Article 9(1) ePrivacy Directive.

[32] Article 29 Working Party (2005). *Opinion 5/2005 on the use of location data with a view to providing value added services*, http://ec.europa.eu/justice/policies/privacy/docs/wpdocs/2005/wp115_en.pdf (accessed on 19/11/2011).

The information should be given in a 'clear, complete and comprehensive'[33] way, focusing on the features of the value added service.

A location based service falling under the scope of Article 9 of the ePrivacy Directive can either be provided directly by the electronic communications operator, or there may be third parties involved, besides the electronic communications operator, who provide the service based on location information that they obtain from the operator. However, only one of the parties that are involved for the provision of the location based service should be responsible to offer the information relating to the processing of location data to the user or the subscriber: the one that determines the means and the purposes for the processing of the data and qualifies as data controller. In practice, this will be the party that is collecting the location data for processing, which in principle will be the provider of the location based service. When this party does not have a direct contact with the user or the subscriber, the information should be provided by the electronic communications operator[34].

The information can be provided in various ways. It can for instance be provided every time the service is used, or in the general terms and conditions for the location based service[35]. In the latter case the service provider should make the information available so that the individuals concerned can consult it at any time and easily, such as via visiting a dedicated website or while using the service, by dialling, for instance, a toll-free number[36]. It is questionable whether the information can be given in the general terms of the contract that is concluded between the subscriber and the operator, if it is clearly highlighted as relating to the provision of the location based service.

Consent

The Article 29 Working Party has taken the position that the consent for the processing of location data other than traffic data for the provision of location based services cannot be given as part of accepting the general terms and conditions for the electronic communications service that is offered[37]. The consent can however be given in the general terms and conditions for the location based service itself, in order to avoid the nuisance from consenting to each transmission of data for the purpose of providing the same value added service[38]. In any case, the U.K. Information Commissioner's Office clarified that "a blanket

[33] Ibid.

[34] Ibid. A practical examination on the issues relating to which entity and in which way should provide the information to the user on the basis of a Location Based Services prototype that was developed within the EU funded PRIME project (https://www.prime-project.eu/) is presented in E. Kosta et al. (2008). "Legal considerations on privacy-enhancing Location Based Services using PRIME technology." *Computer Law and Security Review* 24(2), 139-146.

[35] Article 29 Working Party (2005). *Opinion 5/2005 on the use of location data with a view to providing value added services*, http://ec.europa.eu/justice/policies/privacy/docs/wpdocs/2005/wp115_en.pdf (accessed on 19/11/2011).

[36] Idem.

[37] Article 29 Working Party (2005). *Opinion 5/2005 on the use of location data with a view to providing value added services*, http://ec.europa.eu/justice/policies/privacy/docs/wpdocs/2005/wp115_en.pdf (accessed on 19/11/2011).

[38] J. Eckhart (2008) "Zehnter Teil. Telekommunikationsgesetz" (TKG). In Spindler, G. and Schuster, F. (ed.). *Recht der elektronischen Medien*. Munich: Verlag C.H. Beck, 809-1400; A. Ohlenburg (2004). "Der neue Telekommunikationsdatenschutz - Eine Darstellung von Teil 7 Abschnitt 2 TKG." *Multimedia und* Recht, 431-440 [in German].

'catch all' statement on a bill or a website"[39] would not be sufficient to constitute valid consent, but a specific informed consent should be acquired for each value added service.

When the automatic localisation of the individual is required for the provision of a location based service, the initiation of the process by the individual would amount to consenting to being located, provided that he or she is fully provided with the aforementioned information relating to the processing of his location data[40]. For instance, when an individual can dial a specific number or send an SMS to a dedicated number in order to get information on the weather conditions at his or her location, then the dialling of the number or the sending of the SMS should be considered as consent to the processing of his location data.

The Data Protection Directive in its Article 7 enumerates six grounds that can legitimise the processing of personal data, ranging from the consent of the data subject to the protection of his vital interests and the legitimate interests of the data controller. However, Article 9 of the ePrivacy Directive allows the processing of location data other than traffic data for the provision of a value added service only when the prior consent of the user or the subscriber is obtained. Given the lack of direct communication between the user and the location based service provider, under the current deployment of location based services, this choice of the European legislator can be criticised.

In the context of the provision of information to the individual before obtaining his consent, it has been suggested that when there are various parties involved with the provision of the location based service apart from the electronic communications operator, the information should in principle be offered by the party that is collecting the location information for processing, i.e. by the providers of the location based service. Similarly, the same party should be the one to obtain the consent of the individual concerned[41].

In the majority of the current commercial systems of location based services, the electronic communications operator systematically sends location data of the individual concerned to the providers of the value added service upon their request for the provision of the service, except when this information is automatically created by the terminal equipment of the user[42]. The Article 29 Working Party expressed doubts about these current practices that allow for the identification of the user by the location based service provider, while this is not necessary for the successful provision of the service. Therefore, it suggested that the electronic communications operators should transfer the localisation requests to the third party, the provider of the location based service, in a way that will not allow the identification of the individual by the third party[43]. In such a case, the consent of the individual should be collected by the electronic communications operator, as the individual will remain anonymous to the provider of the location based service[44].

---

[39] United Kingdom Information Commissioner's Office (2006). *Guidance on the Privacy and Electronic Communications (EC Directive) Regulations 2003*, http://www.ico.gov.uk/for_organisations/privacy_and_electronic_communications/the_guide.aspx (accessed on 18/10/2011).

[40] Article 29 Working Party (2005). *Opinion 5/2005 on the use of location data with a view to providing value added services*, http://ec.europa.eu/justice/policies/privacy/docs/wpdocs/2005/wp115_en.pdf (accessed on 19/11/2011).

[41] Ibid.

[42] Ibid.

[43] Ibid.

[44] The EU funded research project PRIME developed such a prototypical implementation of a location based service, which decouples the entities involved in a location based services scenario via a trusted location intermediary: E. Kosta et al. (2008). "Legal considerations on privacy-enhancing Location Based Services using PRIME technology". *Computer Law and Security Review* 24(2), 139-146.

The nature of location based services implies a lack of direct communication between the individual who requests the service on the one hand, and the providers of the publicly available electronic communications service and the location based service on the other hand. Therefore, it is important that the provider of the location based service (or the electronic communications service provider, when this entity is defined as data controller), takes measures in order to ensure that the location data that are going to be processed belong to the same person that is consenting to the processing. To this end, the provider should confirm the subscription to the location based service after the receipt of the consent. This can for instance be realised via sending an SMS to the terminal equipment of the user. If it is rendered necessary, the provider should request a further confirmation of the subscription[45]. The Citizens' Right Directive recognised that the consent of the user, when this is technically feasible and effective, can be expressed via the configuration of the appropriate settings of an application[46]. Although this clarification was made in the context of Article 5(3), which has been presented above, it could be applied in the context of location based services as well.

Time validity of consent

When the location based service requires the localisation of the user on an ongoing basis, it suffices that the consent is provided once before the localisation takes place and after the individual concerned is duly informed about the details relating to it. In addition, the provider should send regular reminders to the terminal equipment of the individual about the localisation[47]. However, the danger lurks that the use of the location based service may become cumbersome in case the users are sent notifications too frequently[48]. The German legislator has concretised the frequency of these reminders in relation to value added services that require the processing of other subscribers or third parties besides the provider of the value added service. In these cases, the service provider has to inform the subscriber via an SMS on the number of the localisations he has realised after maximum the fifth localisation via the location data of the mobile device, unless the subscriber has objected.

---

[45] Article 29 Working Party (2005). *Opinion 5/2005 on the use of location data with a view to providing value added services*, http://ec.europa.eu/justice/policies/privacy/docs/wpdocs/2005/wp115_en.pdf (accessed on 19/11/2011).

[46] Recital 66 Citizens Rights Directive (European Parliament and Council of the European Union (2009). Directive 2009/136/EC of 25 November 2009 amending Directive 2002/22/EC on universal service and users' rights relating to electronic communications networks and services, Directive 2002/58/EC concerning the processing of personal data and the protection of privacy in the electronic communications sector and Regulation (EC) No 2006/2004 on cooperation between national authorities responsible for the enforcement of consumer protection laws, *OJ L* 337, 11.

[47] Article 29 Working Party (2005). *Opinion 5/2005 on the use of location data with a view to providing value added services*, http://ec.europa.eu/justice/policies/privacy/docs/wpdocs/2005/wp115_en.pdf (accessed on 19/11/2011).

[48] M. Gadzheva (2007). "Privacy concerns pertaining to location-based services". *International Journal of Intercultural Information Management* 1(1), 49-57.

Concluding thoughts

It is evident that an enormous amount of information is buried in the data available from persistently tracking people. While it is clear that a one month period is by no means long enough to draw a rich behavioural profile, this study has demonstrated that over relatively short periods of time personal, and in some cases sensitive, information can be revealed. What is also clear is that idiosyncratic behaviour will be difficult to model and profile, and would only come together after vast amounts of time monitoring the individual, if at all. This presents the very real potential for incorrect conclusions being reached based on the data, the impact of which on the individual could be significant. Given these issues, it is certainly in the interests of the individuals involved to make sure that they are aware of how this information is being further exploited by those offering the services they are using.

The processing of location data other than traffic data for the provision of a value added service is specifically regulated in Article 9 of the ePrivacy Directive. Such processing is only allowed when the data are made anonymous or with the consent of the subscriber or the user and to the extent and for the duration of the provision of the service. The choice of the legislator to allow the processing of location data other than traffic data only when the consent of the data subject is acquired can be justified, as the provision of value added services are additional services that are offered for the enhancement of the personal experiences of the users. However, Article 9 has a limited scope as it applies only to location data that indicate the geographic location of the terminal equipment of a user of a publicly available electronic communications service, and to location data that are transmitted via a public communications network. Clearly, the consent of the user will be, thus, required for the processing of his location data for the provision of a location based service that enables him to find the nearest pharmacy via his mobile phone. Would his consent also be required if his location data are transmitted via an RFID network or via a Bluetooth network? In order to answer this question, it is crucial to specify if the said network is public or not.

In case the electronic communications network is not public, then the processing of the location data will remain outside the protective ambit of Article 9 of the ePrivacy Directive. In such a case, the processing will have to be conducted in accordance with the general provision of the Data Protection Directive, which allows the processing of personal data not only when the individual concerned has given his consent (as in Article 9 of the ePrivacy Directive), but on any of the grounds that are mentioned in Article 7 of the Data Protection Directive.

The situation that is created from the narrow scope of application of the Article 9 of the ePrivacy Directive raises difficulties for the providers of a location based service, who will have to carefully examine the data they are processing for the provision of the service, the nature of the service they are offering, as well as the type of network that is used for the provision of the service, in order to conclude whether the acquisition of the consent of the individual concerned is obligatory for the processing of his location data. At the same time, location based services offered via technologies that do not invoke the application of Article 9 of the ePrivacy Directive will not necessarily have to rely on the consent of the individual concerned. Cuijpers and Pekárek have suggested two approaches that would ensure a homogeneous treatment of location data in the deployment of location based services, irrespective of the technologies used. On the one hand, the scope of application of the ePrivacy Directive could be extended in order to cover all types of location based services, or on the other hand, the processing of location data could be removed from the scope of application of the ePrivacy Directive and such processing would be governed by the Data Protection Directive in all types of location based services[49]. The adoption of the latter approach, besides ensuring the uniform treatment of location data in all types of location based services, would allow the processing of location data based on other legitimate grounds. This could be considered as a positive development in the processing of location data, as the validity of the consent of the user or the subscriber in the form that is provided today is questionable, when the users have to accept the processing of their location data under the conditions offered by the provider in order to enjoy the service.

Acknowledgments
Google API Powered by Google – Reproduced with Permission, Map Imagery © 2009 DigitalGlobe, GeoContent, GeoEye, AeroWest, Map Data © 2009 Tele Atlas

[49] C. Cuijpers and M. Pekárek (2011). "The regulation of Location Based Services: Challenges to the European Union Data Protection Regime". *Journal of Location Based Services* 5(4), 223-241.

Geographic Data and Legal Liability Issues

George Cho[1]

Introduction

In the 21st century, one could say that the digital map has fully taken over from its paper predecessor. Marine charts, for example, have been replaced by digital maps displayed on computer monitors. If digital technology is to become the universal means of using geographic data, then this chapter suggests that mapping services provided over the Internet may become a source of new legal liabilities and risks. As the use of geographic data becomes more widespread and as the technology supporting the geographic information systems (GIS) industry matures, it is opportune to reconsider some of the rights, responsibilities and liabilities of both the suppliers and users of such data. All too often one reads about geographic data being blamed for serious accidents or disasters, raising questions about who is responsible for these accidents or disasters. This chapter highlights several legal liability issues that have come to light in the application of GIS and evaluates where the responsibilities may lie for possible damage or harm caused by the use or misuse of geographic data. In addition, it discusses the effect of new dilemmas emerging from the evolving nature of the geographic industry by examining two examples: global positioning systems (GPS) and the use of Web 2.0 technologies.

The chapter begins with a brief survey of emerging issues arising from the use of geographic data and the myriad ways in which legal liabilities may be incurred in the age of the Internet. The survey purports to demonstrate that traditional legal analyses underlying legal liability may have become less appropriate for dealing with geographic data. To appreciate this proposition, early conceptions of the legal framework applicable to geographic data need to be discussed first, and liability risks in the production, use and deployment of geographic data and associated technology need to be assessed. Next, developments in the use of geographic data such as GPS and maps that are stored on remote servers in the distributed 'cloud' environment (so-called cloud sourced), are scrutinized as pointers to future liability risks. The concluding section provides a summary of the recent developments in the use of geographic data and legal liability issues.

A survey of emerging issues

Liability issues with regard to geographic data are not new. A frequent lament heard at official inquiries by safety authorities into accidents and disasters is that faulty maps are to blame for the damage. For example, the U.S. Mine Safety and Health Administration Report of July 2003 cited faulty maps as a cause for the Que Creek Mine disaster in June 2002, where

[1] University of Canberra, George.Cho@canberra.edu.au.

nine miners were trapped underground for four days until they were rescued through a narrow bore hole. The report stated that

> "the primary cause of the water inundation was the use of an undated and uncertified mine map of the Harrison No. 2 mine that did not show the complete and final mine workings. Using this map led to an inaccurate depiction of the Harrison No. 2 Mine workings on the Que Creek#1 mine map as required by the Mine Safety and Health Administration"[2].

The key words that strike the reader are *undated*, *uncertified* and *inaccurate*. These lead to questions about who prepares the so-called *faulty* maps and who certifies them. For geographic data it may appear that one of the myriad ways of discovering answers to these questions is through metadata (data about data), in the form of digital files or the marginal information found on printed maps. However, such extraneous information is not always sufficient to determine where the actual liability lies, as there may be a combination of factors lying at the base of the erroneous data.

In another example, the Norwegian State agency that publishes marine charts and maps was charged in court with responsibility for the fatal grounding of the bulk ship *Rocknes* in 2005 off the west coast of Norway. The State prosecutors claimed that the agency did not offer sufficient information about the hazardous area where the bulker grounded and capsized, losing 18 crew members and dumping a large amount of oil into the sea. However, the charges were later withdrawn because the accident was held to be most likely caused by navigational errors by the ship's crew. The appeal against this decision before the Appeals Selection Committee of the Norwegian Supreme Court was dismissed in May 2011[3].

Many other cases can be found. For instance, police and ambulance services were sent to the wrong location in south-west Sydney, Australia in response to an emergency call due to errors in the maps that were used. An out-of-date map increased the response time as emergency crew was sent to a street with the same name but in a different suburb[4]. In another case, questionable directions given by an online mapping service could have contributed to the death of a person who perished while trying to save his family. The driver stranded in the wilderness in Oregon in the U.S., where Google Maps had suggested a short cut through roads that became dangerous in winter. The driver left his family in the car in order to get help, but never returned[5]. Next, in 2010, a woman sued Google, claiming that the Google map she had used led her to a dangerous Park City roadway without room for pedestrians, where she was struck by a car[6]. In her writ she stated that "Google undertook the duty to exercise reasonable care in providing safe directions to patrons of its Google

---

[2] See Mine Safety and Health Administration (2002). *Report of investigation. Underground Coal Mine Nonfatal Entrapment*, www.msha.gov/quecreek/QueCreekInvestigationReport.pdf (accessed on 16/10/2011); Directions Magazine Staff (2003). "Faulty maps to blame for Que Creek Mine Disaster", http://www.directionsmag.com/article.php?article_id=407 (accessed on 17/10/2011).

[3] Hydro International (2011). "Rocknes Claims Appeal Dismissed", http://www.hydro-international.com/news/id4822-Rocknes_claims_Appeal_Dismissed.html (accessed on 17/11/2011).

[4] ABC News Online (2005). "Old maps see emergency crew sent to wrong suburb", http://www.abc.net.au/news/stories/2005/11/30/1519577.htm (accessed on 17/11/2011).

[5] A. Moses (2006). "Online map linked to fatal turn", http://www.smh.com.au/news/technology/online-map-linked-to-fatal-turn/2006/12/08/1165081119464.html (accessed on 16/11/2011).

[6] C. Smart (2010). "Woman sues Google after Park City accident", http://www.sltrib.com/news/ci_15199273 (accessed on 17/10/2011).

Maps service [but] Google failed to warn the plaintiff of said known dangers". According to the Court, Google did not have a duty of care towards the woman, because it did not have a special relationship with her and only published information to the general public[7]. Another salutary lesson can be derived from the case of an American driver of a rental car who steered onto railway tracks, based on the instructions given by the car's GPS. He got stuck on the tracks, tried unsuccessfully to reverse and finally abandoned the car. The car was hit by an on-coming metro train. The car driver was cited for obstructing the railway crossing[8].

In all the examples mentioned above, questions arise as to who should be held liable for the damage incurred, and what role is played by disclaimers and metadata. How far should the common sense and rational behaviour of the persons involved be taken into account in adjudicating on the geographic data provider's responsibility? This is particularly critical in the current fast-paced information society. Internet mapping services, such as Google Earth or MapQuest, may include a constant stream of updated maps that are displayed to the end user in real time. These maps are the result of routing algorithms that work in the background in response to incoming data feeds from a database as well as to requests for re-routings from users. Up-to-the-minute detours and re-routing algorithms take into account major bottlenecks, accidents and other impediments on the road. Such real time re-routing is not just a computational nightmare, the resulting fastest route or shortest distance solution could also be the most dangerous. The suggested route might be the shortest in drive time or length, but the track may wend its way through school areas and residential suburbs or through commercially zoned areas that pose other hazards. In such solutions the need for human intervention becomes much greater and the assignment of liability becomes very difficult[9].

The liability risks posed by the computational activities mentioned above suggest that when the volume of map requests and map displays *pushed* to end users increases, the potential for making mistakes, getting routes wrong and other similar problems may become even greater. It is thus absolutely vital that the geographic data that are used in these applications are the very best available in terms of timeliness and accuracy. However, the elimination of errors may by no means be achievable, especially when the data are obtained from many parties. Errors may creep into the geographic database either through multiple handling or through the integration of diverse data sets that may be variable in time, accuracy and resolution. Constant checking and benchmarking with quality standards may be required to ensure accurate, error-free data[10]. The question arises whether conventional standards of care can still be applied in this new digital environment, and whether conventional legal liability theory is sufficiently robust to grapple with technologies of the 21st century. It is to these conundrums that we turn in the next section.

---

[7] V. Balasubramani (2011). "Injured Pedestrian Can't Sue Google for Providing Faulty Map Information – Rosenberg v. Harwood", http://blog.ericgoldman.org/archives/2011/06/injured_pedestr.htm (accessed on 17/10/2011).

[8] N. Neroulias (2008). "Driver cited in Bedford train-car crash caused by GPS mishap", http://www.dslreports.com/forum/remark,19738907 (accessed on 17/10/2011).

[9] R. Konrad (2007). "Online maps updated after highway disaster", http://www.usatoday.com/tech/webguide/internetlife/2007-05-03-online-maps_N.htm (accessed on 17/10/2011).

[10] It is reported that TeleAtlas has 5.5 million U.S. streets in its database and about 3 million changes are made to its maps each month. See J. Saranow (2008). "Steered Wrong: Drivers Trust GPS Even to a Fault", http://online.wsj.com/article/SB120578983252543135.html (accessed on 17/10/2011).

Conceptions of the legal framework on liability for geographic data

This section deals with possible tort and contract liability with regard to geographic data. After setting out the main principles of both types of liability, it examines some court cases from the United States and discusses the challenges that arise with the development of geographic data and GIS technology.

Tort liability

Laws regarding liability are an important part of the geographic industry and may have a considerable impact on government agencies and private organisations that use, sell or buy geographic data, geographic software and other products. Exposure to liability may occur in several ways, e.g. by the provision of inaccurate data or the misuse of accurate information; by the sale of data without warnings about data errors, what the information represents or what computations have been made to produce the data; or by the marketing of geographic data, GIS products and geographic software[11]. In each case, the factual context is critical in determining whether or not there is liability and where the liability may lie.

The concept of liability derives from the French word *tort*, which means *wrong*. This wrong can be a private or public wrong or an injury to a person or property. A *private tort* is an offence against an individual, whereas a *public wrong* or crime is an offence against the community at large.

Under common law principles, tort liability may arise from a number of sources including negligence, fraud, and product liability. *Negligence* is conduct that breaches a duty of care for the protection of others against an unreasonable risk of harm. A person can be held liable for conduct inconsistent with that of a *reasonable person* causing foreseeable harm or damage. Where all these requirements are present it is said that one may be held legally liable. The duty of care that is required may differ. For instance, as the producers of geographic data are traditionally certified surveyors and cartographers, their duty of care may be greater than it would be for ordinary citizens. These professionals work under agreed industry codes of practice and undergo rigorous training and certification. Such processes ensure a requisite standard of care in the preparation of products and in the delivery of professional services.

In the case of maps, liability through negligence may arise from errors in the maps, through poor design, or the use of maps in inappropriate ways or in a manner unintended by the map designers. In attempting to apportion liability, courts will have to consider the different steps taken to produce the maps. One would have to find out how the data were entered into a database, what sort of checks were undertaken to ensure that the data were error-free and what standards of data quality were used. Next, the inquiry will also have to ask about the design and the cartographic techniques that were used and whether these were appropriate for the purposes at hand. Other considerations may include data design rules, personnel training and data compatibility checking procedures in order to ensure that the in-house rules and criteria have been followed.

---

[11] See G. Cho (2005). *Geographic Information Science: Mastering the Legal Issues*. Chichester: John Wiley & Sons.

A second kind of tort liability includes variations of *fraud*, for example fraudulent misrepresentation and inducement of contract. The former may take the form of misrepresentation of fact or opinion to persuade someone to act or not act in a certain manner knowing that that may result in some benefit to the person making the misrepresentation. Inducement to contract occurs when a party to contract negotiations knowingly provides wrong information designed to induce the other to agree to the contract.

The third type of tort arises from *product liability*, where defective and unsafe products result in harm to people and loss and damage to property. Such laws, sometimes described as strict or statutory liability laws, are designed to ensure that products reaching the market meet minimum industry quality standards in order to ensure consumer protection. The relationship between product liability and liability based on negligence with regard to geographic data and products depends on whether GIS is considered a service or a product. The liabilities attached to these are significantly different. GIS as a service relates to the production of high quality maps and written reports derived from geographic data. If, on the other hand, one were in the business of gathering geographic data and selling them without analysis and further processing, then such an activity might result in GIS products (data only). So in the case of providing data, for example, if there are errors in the data that are not made known to the user the supplier may potentially be liable for any injury or loss as a result of their use. Next, where GIS is the supply of geographic data and is found to be not 'fit for [the] purpose' for which it was intended, there may be product liability.

Contractual agreements and liability

Liability for breaching mutually agreed contractual conditions can also play a role in the relationship between producers and users of geographic data. Legal liabilities may arise through contract, but they may also be limited in the contract through warranties and disclaimers with regard to the quality of the data or their fitness for purpose. Traditional principles of contract law cover issues of whether there is a legally enforceable agreement to do something, e.g. to produce a result, to write a report, or to process raw data into useful information[12]. Another facet of contract law relates to the licensing agreements for bespoke computer software, the use of the computer software, or simply the use of geographic data. Terms may relate to the assignment of intellectual property rights (IPR), the ownership of the bespoke software and escrow arrangements for the source code in the event of the collapse of the software company, and other incidents that prevent the fulfilment of the terms of the contract[13].

The key ingredients of a contract are the conditions and warranties. These conditions and warranties must be clearly set out at the initial stage before the contract is executed. Some agreements also give details of the limits to liability in exclusion clauses that exempt or exclude certain types of liabilities and place a cap on financial liabilities. *Force majeure* clauses for losses arising from extreme events may also be written into the contract. Problems can occur when the conditions and warranties are unclear and where certain

[12] Ibid.

[13] In an escrow arrangement, for example, the source code of computer software may be delivered to a trusted 'stranger' who is not party to the contract. This party takes custody of the software code to ensure that certain conditions are performed or met. In the written agreement the purpose of an escrow is explained to facilitate access to the software code in cases where the software developer or company becomes unable to service the program.

types of liability are not excluded or where the limits to financial liability have not been predetermined and stated.

With regard to contracts that license the use of geographic data, there is a view that while a data owner may have agreed to share data and has waived ownership of IPR, the legal liability may re-emerge when data are misused. Pomfret suggests that while a data licensee may agree to certain terms for the use of the data supplied, the data provider has not given up all the responsibilities of that ownership[14]. Such terms may include the use of the data in a way that complies with applicable law, an acknowledgment that the data would be fit for specific purposes, and/or an agreement not to sue the licensor if there are errors in the data set. Yet, even with such contractual provisions, the data owner may still have to be concerned about legal liability if someone is injured as a result of the way the data are used. Thus, it may be uncertain when the data owner is relieved of all responsibility and liability. In addition, if the data holder is a public body, it may be subject to particular duties to provide data. For instance, in the U.S. under federal law such as the Freedom of Information Act and the Open Records regime, anyone can request geographic data and the data are supplied on an *as is* basis with all care taken in their preparation and no responsibility attaching to their subsequent use or misuse[15]. On the other hand, in Australia and New Zealand, the *custodianship* regime requires that the government agency has the responsibility not only of keeping custody of public data and information but also their subsequent use either in-house or by other agencies and bodies. Hence, there is an implied warranty that the information may be relied upon[16].

Finally, contractual issues cannot be considered in isolation as there are interactions between the laws of contract and tort, and between these laws and statutory obligations. The contract-tort nexus may arise in each contract where there is an agreement to do something. If losses or damages are incurred in the carrying out of those tasks and these are of a tortuous kind, the relationship between the contractual liability and the tort liability becomes difficult. Also, there may be regulations that include specific statutory obligations with regard to e.g. occupational health and safety apart from contract liability and tort liability. If these regulations were infringed, statutory liability would also arise next to the general contract or tort liability.

Examples of court cases

In the U.S., a number of cases involving maps have been litigated in the course of the second half of the 20th century, each resting on particular factual situations. U.S. courts have consistently viewed maps and charts as 'practical' – rather than 'creative' – works the purpose of which is to guide. This distinction is critical because in the U.S. aeronautical and navigational charts based on geographic data have been held to be *products* and thus subject to product

[14] K. Pomfret (2010). "Onus of the owner". *GIS Development* 14(6), 38-40.

[15] H.J. Onsrud (1995). "Role of Law in impeding and facilitating the sharing of Geographic Information". In Onsrud, H.J. and Rushton, G. (ed.). *Sharing Geographic Information*. New Brunswick: Rutgers, 292-306.

[16] Australian Land Information Council (1990). *Data See Custodianship / Trusteeship, Issues in Land Information Management*. Belconnen: ALIC.

liability laws[17]. Creative works could include art works, the writing of poetry and other written works, and as such there is a different kind of copyright law that will apply.

An example of the distinction is found in *Brocklesby v United States*[18], where the issue was whether an air navigation chart is a 'product'. In this case, an aircraft crashed into a mountain and an action was brought against the chart maker for publishing a defective product. The charts were designed for use by aircraft pilots who made immediate decisions based on the information they contained. As such the standard of care in preparing these maps was considered critical because of the element of reliance that was placed on the use of them. The court held that the chart was a 'product in a defective condition unreasonably dangerous to the user' within the meaning of section 402A of the *Restatement (Second) of Torts (1965)*, and that the chart maker should be held liable.

Such a duty of care was also illustrated in *Brocklesby v Jeppesen*[19]. In this case an aircraft pilot using a chart produced by Jeppesen crashed into a mountain, killing the entire crew. Even though the chart publisher had used information obtained from the U.S. Federal Aviation Administration (which could arguably be considered as 'official' or 'of sufficient quality') in the chart, the court ruled that there was negligence, a breach of warranty and strict liability on the basis that the chart maker was under a duty of care to test its products and to warn of any inherent dangers the use of the charts entailed.

Errors and omissions in represented locations causing damage have also been litigated in various other cases before U.S. courts. The cases include omissions, inaccuracy and negligent representations that amounted to negligence and a breach of statutory duty. In *Indian Towing Co. v United States*[20], the U.S. government was held to have negligently failed to maintain a lighthouse marked on nautical charts. Similarly in *Remiga v United States*[21], the federal government was held to have inaccurately and negligently depicted the location of a broadcasting tower on an aeronautical chart, which contributed to mistakes and resulted in fatalities in an aeroplane accident. In *The City of Tarpon Springs v Garrigan*[22], plaintiffs suffered loss when it was negligently represented to them that they were not building in a flood plain and they were subsequently unable to obtain insurance against floods.

Perversely, it may be possible that all the information provided may be correct, but has been incorrectly used. For instance, this issue was raised in *Aetna Casualty & Security Co. v Jefferson & Co*[23]. In this case liability was found because the form of representation of the geographic data was either confusing or inappropriate for the user's purposes. The product in question was an aeronautical chart that depicted an instrument approach to an airport. The display was in a graphical form with two perspectives – one from above, a plan view, and the other from the side, a profile view. The court found that the chart was defective not in the accuracy of the data but in the manner of graphical representation. The side or profile view appeared to be on the same scale as the plan view when in fact there was a difference between the two by a factor of five. The crew of the crashed aircraft was misled in part by the method used in the graphical representation. The publishers of the chart were found to

[17] See N.J. Obermeyer and J.K. Pinto (2008). *Managing Geographic Information Systems. 2nd ed.* New York: The Guildford Press.
[18] 753 F.2d (9th Cir. 1985).
[19] 767 F.2d 1288 (9th Cir. 1985); *cert. den.* 474 US 1101 (1986).
[20] 350 US 61 (1955).
[21] 448 F. Supp. 45 (1978) (WD Mich.).
[22] 510 So. 2d. 1198 (1987).
[23] 642 F. 2d. 339 (9th Cir.) (1981).

be at fault since professionals must be able to rely on the accuracy of the information in the chart. However, the crew were also partly at fault, and contributed to the negligence, as they were expected to use their professional judgement in evaluating information and not to rely blindly on what was being shown on the charts.

Discussion

The examples illustrated above suggest that with the advent of automated cartography and intelligent GIS products, there may be many more complicated cases that can potentially arise and that may be almost impossible to resolve. In addition, the risks may be even greater for the ever increasing number of novice users of geographic data that do not have an intimate knowledge of the data or the way they are represented. Details such as how and where the data were collected, what manipulations have been performed on the data, and what assumptions and omissions are at the background of the final result may become increasingly blurred the more the user is separated in time and circumstance from the creator(s) of the geographic data. Without a good knowledge of the origins of the data, users may be more likely to make innocent mistakes that can unfortunately result in collateral damage and unintended consequences.

The cases discussed above point to the difficulty in trying to establish a requisite standard of care in GIS, whether for a product, a service or a combination of the two. These difficulties may also include liabilities arising from third party involvement. For example, geographic data sold or produced by one agency are used by another party who makes additions to the database, but in doing so inadvertently introduces error. When someone else then uses these erroneous data, the person to be blamed in this *data chain* may be difficult to trace. In cases where there is joint liability, working out the proportional share of the blame is the next challenge. In addition, there may be vicarious liability – in the sense that the liability falls on one person as a result of the action of another. A common example is the liability of an employer for the acts and omissions of employees.

The issues of liability in tort and contract suggest that one would need to be very careful in the collection, use, ownership, sharing and sale of geographic data and information[24]. However carefully the legal rules have been considered and whatever legal risk management plans have been put in place, there can still be unexpected pitfalls. The instances where legal liability may arise are not predefined, and liability is evolving as quickly as the technology itself. In this respect, the legal liability for distributing and using geographic data is thought to be one of the most explosive issues in the emergence of GIS technology[25].

However, the litigation that has come to the courts has been sporadic and fragmented, with the consequence that little advance has been made in either the development of the law or the understanding of the application of the law. The lack of case law may be due to the reluctance of the parties to go to court and their preference for settling out of court, based on the wish to avoid adverse publicity, limit compensation and/or litigation costs, or minimise further risks of litigation. In addition, a lot of uncertainty exists not only with reference to the technology itself but also with regard to the agreed technical standards and

[24] G. Cho (2005). *Geographic Information Science: Mastering the Legal Issues*. Chichester: John Wiley & Sons.

[25] E.F. Epstein (1991). "Legal aspects of geographic information systems". In Maguire, D.J., Goodchild, M.F. and Rhind D.W. (ed.). *Geographic Information Systems: Principles and Applications*. London: Longman, 489-502.

established professional industry codes to be applied, hindering the possibility of basing liability principles on these standards and codes[26]. While it remains imperative that liability for loss and damage may be ascribed to someone or some agency in order to ensure that compensation and repair can be effected, it becomes increasingly difficult to determine who should be blamed.

The question that needs to be asked next is whether there are *new* liability risks that are inherent in the new models of collection, use and dissemination of geographic data. In the next section, this question will be addressed with regard to GPS technology and its myriad applications and risks arising from the use of online maps and user generated content.

Geographic data and new liability risks

GPS technology

Earlier in this chapter, a number of examples were given of situations in which reliance on maps and instruments could cause parties to be held responsible for accidents and damage. In some of these cases the defence might be that 'the GPS instrument instructed me to do so'. Such defence may seem counter-intuitive, lacking common sense, or even laughable – but is it? Is there a case against the GPS provider rather than the user? A number of aspects have to be taken into account in a case of inaccurate directions provided by a GPS device.

A first example is the use of GPS for navigation purposes. Sometimes the positional readings given to a GPS user may be incorrect and the user may not have noticed because of his or her over-reliance on the use of technology that has replaced ordinary common sense[27]. On the one hand, the user could argue that he has a contract for the navigation service and that he or she is therefore entitled to rely on the accuracy of the provision of that service. However, in contractual terms it is likely that the user would have to accept a disclaimer by the maker of the GPS device[28]. The contract would also require the user to ensure that the device is used in a proper and safe manner despite what the GPS device itself may instruct the user to do. A relevant question here is whether this is an enforceable contractual provision which the parties have mutually agreed to or whether it could be considered an unenforceable *adhesion contract*. An adhesion contract is a type of contract between two parties in which one side has all the bargaining power and uses it to write the contract primarily to his or her advantage. A well-known example is standard contracts for goods and services on a *take it or leave it* basis, e.g. for computer software[29]. Such adhesion or standard contracts are not always valid. For example, in the European Directive on Unfair Terms of Consumer Contracts, terms that are unfair and result in a significant imbalance of

[26] See the discussion on the MAPPS lawsuit in D. DiBiase et al. (2008). "Scoping GIS for education: Making sense of academic and practitioner perspectives". *Geography Compass* 2(5), 1506-1528.

[27] R.G. Satter (2006). "GPS sends ambulance 200 miles off-route", http://www.usatoday.com/tech/products/gear/2006-12-07-gps-fault_x.htm (accessed on 17/10/2011); J. Banard (2009). "Couple stranded three days after GPS leads them astray", http://www.usatoday.com/tech/news/2009-12-28-GPS-stranding_N.htm?csp=Tech (accessed on 17/10/2011).

[28] See an example of a disclaimer in E. Boucaut (1992). "Geographic Information Systems for environmental planning and impact assessment". *LIS News* 27(1), 10-12.

[29] See C. Mak (2008). *Fundamental rights in European contract law: A comparison of the impact of fundamental rights on contractual relationships in Germany, the Netherlands, Italy, and England*. Alphen aan den Rijn: Kluwer Law International.

obligations between the parties to the detriment of the consumer are invalid and hence unenforceable[30].

From its point of view, a GPS service provider could argue that the user must take much more care and responsibility in deploying GPS devices. Even with the assistance of a GPS, drivers must pay attention to the road and not pass the blame on the GPS provider when accidents happen. However, this is not always self-evident. For instance, when using GPS in unfamiliar territory and on being given explicit instructions by the GPS receiver, a user must make instantaneous decisions. Under those circumstances it may be reasonable for a user to rely wholly on the GPS unit. Aircraft pilots have no choice but to rely completely on GPS instruments, apart from the audio commands given from a control tower[31]. Sometimes GPS instruments on board aircraft may default to a *dead reckoning*[32] or simulator mode. For instance, such a malfunction or default position was cited as one of the causes of a fatal air crash near Benalla, Victoria in 2004. The Australian Transport Safety Bureau's investigation found that the GPS unit would have needed manual inputs for it to fly in a dead reckoning mode[33]. This case shows that the pilots should have taken care that the instruments were working as expected and, where there were doubts, to intervene manually and take corrective action. However, the issue of who is liable has yet to be unequivocally resolved.

A second use of GPS relates to the measuring of distances and finding positions in geographical space, mostly deployed in cadastral surveys. In Australia, the legal status of such measurements from GPS is uncertain because of the issue of legal traceability. Legal traceability involves the ability to reduce any measurement to a legally defined standard whereby it can be related to stated references, usually national or international, through an unbroken chain of comparisons all having stated uncertainties. For example, the fact that location *A* is 20 metres from location *B* is legally traceable if the distance is measured by an instrument that has been calibrated against a standard length that is a legally defined entity held by the Australian National Measurement Laboratory. The challenge in using a GPS is that it is not possible to compare measurements to known standards because the measurements are the result of computation by a digital processor – the *unbroken chain* criterion has not been satisfied. In addition, the GPS instrument may induce variable errors. The accuracy of the measurement can differ between 1 and 20 metres depending on the quality of the GPS receiver. Finally, errors may also stem from the GPS's satellite triangulation. GPS readings rely on bouncing signals off three or four satellites simultaneously for triangulation purposes to find relative positions. Manufacturers of GPS use either one or a combination of these systems to provide positional and timing data. It may be contemplated that there is a slim possibility of a failure of the satellites. Errors may be introduced through such satellite malfunction, variations in transmission strengths and adverse environmental conditions. Sometimes the shutdown of a tracking system network may require the users to seek alternative providers.

New liability issues in the use of GPS devices may also arise in situations where such devices are used to invade on an individual's privacy[34]. The issue of location privacy is growing

---

[30] Council (1993). Directive 93/13/EEC of 5 April 1993 on Unfair Terms in Consumer Contracts, *OJ L* 95, 29.

[31] See this discussion in E.J. Sinrod (2008). "Is GPS liability next?", http://www.news.com/Is-GPS-Liability-next/2010-1033_3-6226346.html (accessed on 17/10/2011).

[32] Dead reckoning is the process of estimating one's current position based upon a previously determined position or fix, and advancing that position based upon known or estimated speeds over elapsed time and course.

[33] S. Creedy (2008). "Faulty GPS units lethal for pilots. Dick Smith tells transport minister", http://theaustraliannie.newspaperdirect.com/epaper/viewer.aspx (accessed on 17/10/2011).

[34] See the contributions of E. Kosta & M.N. Gasson, and K. Pomfret about location data and privacy in this book.

and may feature prominently, especially where hand-held devices such as 3G-enabled GPS, iPhones and iPads provide positional information in the form of geotags in social networks. Geotag information may be given out either consciously or inadvertently[35]. Thus, location based services included in social networks such as MySpace, Foursquare, Gowalla, Google Buzz, Facebook Places and Brightkite may be forced to address the issue of liability, either because their own activities are considered an invasion of privacy, or because their users infringe the right to privacy of other individuals. The extent of liability falling on the providers of these location-based social network services is currently unresolved territory and requires further research.

Next to the dangers for privacy in the social network context, the use of GPS devices in surveillance activities may also endanger the privacy of the individual. Regulations, e.g. the Australian *Surveillance Devices Act* 2004 (Cth) and *Occupational, Health and Safety Act* 1991 (Cth), may prohibit the installation of surveillance devices in cars and trucks to check the vehicle's location, speed and security. For instance, based on this legislation, Telstra Australia was prohibited from installing GPS devices in its corporate vehicle fleet, because there was no genuine consent of its employees and there was a breach of statutory provisions[36]. However, while surveillance by private parties may often be limited by law, this is much less the case with surveillance by e.g. police authorities. The growing risks of GPS surveillance by the State and the possible liability of the State for invasion of the individual's privacy will have to be further researched in the future.

In summary, it can be stated that liability risks in the use of GPS devices loom large because of the dynamic availability of geographic data through such devices. GPS have the unique feature of plotting waypoints to tell users where they have been but also where they are in real time. The GPS device represents a convergence of advanced mapping technology that couples satellite tracking and wireless communications. Additionally, the positional data that are collected may seem ephemeral in nature when compared with hardcopy paper maps. The apportioning of liability risks is thus complex. For example, with GPS readings, positioning data are received by the GPS instrument in 'real' time because it is constantly updated as the satellites sending the data and the receiving unit can be changing position all the time. However, looking at a navigational paper chart positional markers are laid out on the map and can be read off the map using geographical coordinates. The error that may creep in here is the user reading off the incorrect coordinates.

Geographic data in the Web 2.0 environment

The Web 2.0 environment has revolutionised the way data and information are created, disseminated and consumed by users. It has also provided opportunities to monetize and re-purpose content[37]. At the same time such developments have created new challenges for users and producers in the publication and dissemination of user generated content. These creators may be liable for the content they create and upload onto the Internet. In addition,

---

[35] S. Canning (2010). "Geo-tagging tweets a great tip-off for burglars", http://www.theaustralian.com.au/media/geo-tagging-tweets-a-great-tip-off-for-burglars/story-e6frg996-1225890465121 (accessed on 17/10/2011).

[36] E. Hannan (2007). "Telstra faces spying probe", http://www.theaustralian.com.au/australian-it/telstra-faces-spying-probe/story-e6frgalx-1111114067790 (accessed on 17/10/2011).

[37] See for example http://www.drostdesigns.com/web-content-how-to-repurpose-info-products/ (accessed on 17/10/2011) and the economics of digital content in http://paidcontent.org (accessed on 17/10/2011).

Internet content hosts may also be liable as operators and owners of websites from where content is stored and accessed, as are Internet service providers[38]. Many of these Internet intermediaries are potential targets because of their identities, profiles, reputations and brands. They may also be subject to regulatory regimes and duty of care obligations[39].

A considerable part of user generated content has a geographic component and can therefore be considered geographic data. The advent of Web 2.0 combined with the variety of ways in which geographic data and location intelligent tools are used requires a thorough review of the risks of these new evolutions and the consequent liability issues.

For instance, the use of online maps is an example of where the Web 2.0 environment has prospered. Online users are equally active producers, distributors and consumers of content on the Internet. The current trend of *mashups* of user generated data with Google or other maps comes to mind. A mashup is a map in which one uses and combines data and presentations from two or more sources to create new maps or services[40]. These maps or services are communicated to a web audience which may reside across the globe and transcend different legal jurisdictions and cultures.

This also means that cross-jurisdictional issues come into play, and it has to be determined which law is applicable to a particular case and which court has jurisdiction to adjudicate on the case. For instance, in a landmark case from the Australian High Court, it was determined that Australian jurisdiction was triggered by the publication of infringing material on the Internet. Such publication occurs when material is posted on to a web server by a user, and again when downloaded by a third party from the web server. Publication is complete at the time of the download[41]. So how does one apply this to the use of maps that are provided online and the additions that users make to such maps in their mashups? And do courts in other jurisdictions use the same criteria to determine the applicable law and competent court?

Managing the liability risks for user generated content and geographic data can be difficult, but steps may be taken to mitigate the potential liability exposures. Liability may be avoided by deciding in the first place whether or not to publish the user generated geographic data; by bearing the costs of ensuring error-free data by quality checks and update procedures; or by expressing warranty statements about data accuracy as well as other contractual and licensing considerations. However, the validity and sufficiency of these mitigating measures need to be assessed in each individual case.

An example of user generated geographic data projects is OpenStreetMap, which involves the creation of free geographic data such as street maps and the provision of these data to anyone who wants to use the data[42]. The project puts an innovative slant on the traditional map and claims to be suitable for use by anyone – GIS professionals, cartographers, marketers, graphic designers and others. It is similar to a *wiki* or to *Wikipedia*, where the website allows the easy creation or editing of a number of interlinked web pages via a web browser using simple editing tools. However, this creates a risk of lack of accuracy

[38] J. Dwyer and M. Lloyd-Jones (2010). "Internet – The Gutnick decision and its implications for Australian trade mark law", http://www.findlaw.com.au/articles/1586/internet---the-gutnick-decision-and-its-implicatio.aspx (accessed on 17/10/2011).

[39] Minter Ellison (2010). "Riding the Web 2.0 wave – limiting liability for user generated content", http://www.minterellison.com/nl_201003_tmtd/ (accessed on 17/10/2011).

[40] See http://en.wikipedia.org/wiki/Mashup_(web_application_hybrid) (accessed on 17/10/2011).

[41] Dow Jones & Coy Inc. v Gutnick (2002) 210 CR 575; HCA 56.

[42] http://www.openstreetmap.org/ (accessed on 17/10/2011).

and of legal liability – the latter remaining as yet undefined and residing in unchartered territory.

The enthusiasm for services such as Google Earth and Google Maps, the possibility of creating mash-ups on these platforms, and the easy availability of geographic data is not shared world-wide. For example, China requires online map providers to operate under a licence issued by the State Bureau of Surveying and Mapping. In addition, online maps are expected to be hosted on servers based in China. The State Bureau of Surveying and Mapping is concerned about the unauthorized leakage of sensitive government data, such as military information and State secrets, within online maps. The recent move of Google search services offices from mainland China to Hong Kong in June 2010 suggests in this perspective that it may have to stop including maps of China in its Google Map services. The impact on crowd sourced maps is also evident, as many user generated maps will have to be removed under these regulations, which the Chinese government calls an 'effort to regulate the quality of online maps and to promote an *orderly development* of Chinese map services'[43]. For instance, a Japanese national received an administrative punishment because he took measurements inside Longuyan in China and included these in a map that he published[44]. The Chinese approach, while not new in its policy of Internet censorship, is a reminder of the possible exposure of any producer and user of geographic data to regulatory liability in the digital age.

Conclusions

The brief survey of emerging liability issues made in this chapter suggests that the printed map in the 21st century has almost become a relic of the past and is being replaced by digital geographic data products. This entails that mapping services provided over the Internet may become the source of new legal liabilities and risks, under both contract and tort law. In the geographic context, exposure to legal liability hinges on accuracy, error, representation and breach of licensing issues, or the failure to fulfil statutory duties. In many cases presented here, breach of warranties and strict liability are of great concern, as are errors of omission and negligent representations. In addition, the question of the contributory negligence of the user in the misapplication and use of maps and data has also arisen. Here professional standards of both practitioners and suppliers of geographic products play an important role.

When digital cartographic products and automated processes are used, it can be imagined that complex problems may arise that may not have been encountered previously in the paper era. The question is asked: *what is the requisite standard of care for geographic data, and GIS, whether it is a service, a system, a product or a combination of all of these*? Where an instrument such as a GPS device is involved, the answer may yet be undetermined and could be indeterminate. Requisite care and due diligence in the use of these instruments should be utmost in any user's mind, especially where they are used in professional contexts. Through time, increased litigation in the courts may help to develop case material. Attempts

[43] V. Ho (2010). "China's regulations to weed out user-generated maps", http://www.zdnetasia.com/China's-regulations-to-weed-out-user-generated-maps-62200615.htm (accessed on 17/10/2011).

[44] M. Melanson (2010). "China goes after 'illegal' online maps", http://www.readwriteweb.com/archives/china_goes_after_illegal_online_maps.php (accessed on 17/10/2011).

to codify the law in this area will produce an identification of the gaps in the law and practice, especially where cross-jurisdictional cases come to the forefront.

Next to GPS devices, the Web 2.0 environment has also begun to challenge previously held conceptions of rights and responsibilities, due to the facilitation of the creation of new content by any user – in this case mash-ups of data and maps – that is broadcast to anyone with access to the Internet and crosses boundaries and legal jurisdictions. Will such democratization of the geographic industry also create a new legal paradigm where the providers have no responsibility and the user must take extreme care in the use of the products? Or will all the risks associated with the use of geographic data have disappeared as intelligent technology becomes capable of delivering the geographic products virtually error-free? These legal questions need to be addressed as the costs and potential damage brought about by location based services and location aware technologies can be huge. Initially the answers to these questions should lie with the application of traditional law to landmark cases litigated in the courts. At the same time new laws and regulations may need to be developed, and the industry may have to adapt to the new digital mapping environment, e.g. through *soft touch* industry mediated regulation, perhaps subliminal social engineering controls through software, or industry codes of conduct. Ultimately, the end result is one where the protection of the user is balanced with reasonable requirements from the suppliers.

High Resolution Data Dissemination against the Background of National Security in India

Malay Adhikari[1]

Introduction

In recent years, geospatial technology[2] has achieved the technological proficiency to provide us with accurate and efficient data through high resolution (HR) imageries. Generally these HR imageries are accurate up to a ground spatial resolution of less than two metres[3]. The number of satellites in outer space is gradually increasing day by day and the sensors used in these satellites are continuously updated to capture data at an ever higher resolution. Countries across the world, including India, want to use this technology to improve their society[4] and the quality of life of their citizens.

In India, the *Citizens' Charter* of the Indian Space Research Organisation (ISRO) provides that the Department of Space, which is the highest policy making body in the Government regarding any space science and technology related matter, has an obligation to provide satellite imagery for developmental purposes to the Government or any other organisation[5]. The Government of India is currently examining the possibilities for disseminating HR data in the public domain. As early as in 2005 the Government enacted the Right to Information Act[6] (RTI Act), through which every citizen has the right to ask for HR data from public bodies, and can challenge decisions refusing his or her request for such type of data. However, even though the RTI Act promotes the availability of HR data, the government is concerned about the effect on national security, and wants to limit the dissemination of HR data[7]. The conflict between these two interests will be discussed in the following sections.

Importance of high resolution data

The importance of HR data is increasing all over the world with the development of geospatial technology. It is becoming clear that the applications of these technologies and the ensuing

---

[1] Centre for International Legal Studies, School of International Studies, Jawaharlal Nehru University, adhikari69@ gmail.com.

[2] See for more details B. Bhatta (2010). "Geographic(al) Information System, GIScience, Geomatics, Geoinformatics, Geoinformation Technology and Geospatial Technology", http://www.bbhatta.info/bb_blog/index.php?entry=entry100226-013233 (accessed on 22/2/2011).

[3] Fomtchenko, M.M. and Movlyav, A.S. (1997). "High Resolution Remote Sensing: New Aspects and Problems". *Proceedings of the Colloquium of Outer Space*. Washington: American Institute of Aeronautics & Astronautics, 85- 93.

[4] S.K. Das (2007). *Touching Lives- The Little Known Triumphs of the Indian Space Programme. 1st ed.* New York: Black Classics.

[5] ISRO (s.d.). *"Citizens Charter"*, http://www.isro.org/scripts/citizencharter.aspx (accessed on 26/02/2011). See also ISRO (2011). *Remote Sensing Data Policy*, http://www.isro.gov.in/news/pdf/RSDP-2011.pdf (accessed on 05/07/2011).

[6] See http://righttoinformation.gov.in/rti-act.pdf (accessed on 22/2/2011).

[7] R. Ramachandran (2000). "Public Access to Indian Geographical Data". *Current Science* 79(4), 450-467, www.iisc.ernet.in/currsci/aug252000/ramachandra.pdf (accessed on 22/10/2010).

geographic information management are of strategic importance[8]. People from different disciplines, such as environment scientists, lawyers, academics, oceanographers, cartographers, medical researchers, engineers and architects, are increasingly incorporating HR data into their professional activities[9]. Also in India, demands for HR data are rising[10]. Some of these demands have been met recently by the amendment of the Remote Sensing Data Policy (RSDP) in 2011. While previously only data with a resolution of up to 5.8 metres were disseminated without restriction, this is now also the case for data with up to 1 metre resolution[11].

However, the growing use of HR data also raises some concerns with regard to the possible lack of control over the data and how they are spreading. This can be demonstrated by an analogy: traditionally, photographs of a person were often taken by ordinary cameras without the person's knowledge (e.g. in a tourist venue or at a cultural event). Generally, this did not cause any problems. However, if the same photograph is taken by a highly sophisticated digital camera and it can be disseminated broadly without losing quality, the person involved may become more worried. The person's physical identity is more exposed because the technology of the digital camera is better than that of the analogue one; the image can be disseminated an unlimited number of times without losing quality; and anybody else can easily identify the person and take advantage of such photography for harmful activities. Something akin to this is what is happening in the remote sensing image arena. With the increasing availability of HR images, nations are starting to worry (just like the individual mentioned above) about their national security as well as the privacy of the individual citizen, as they feel that they have no control over the imaging of information on their territories[12].

How space law deals with HR data

Background

The main provisions of space law date from a time when the capacity of sensors and satellites was much less. Thus they do not provide an immediate answer to the possible dangers HR data may bring to national security or privacy, simply because such conflicts could not be conceived of at that time. However, these provisions of space law may be the only regulations in the international arena that can provide a solution.

The main space law provisions are included in the *Treaty on Principles Governing the Activities of States in the Exploration and Use of Outer Space, including the Moon and Other*

[8] P. Cheung (2009). "Opening statement". Presentation at 18th session of the United Nations regional cartographic conference for Asia and the Pacific, Bangkok, http://unstats.un.org/unsd/methods/cartog/Asia_and_Pacific/18/OpeningStatement-PC.pdf (accessed on 22/10/2010).

[9] C. Potsiou et al. (2010). "Rapid Urbanization and Mega Cities: The Need for Spatial Information Management". *Conference papers of XXIV FIG International Congress*, http://www.fig.net/pub/monthly_articles/march_2010/march_2010_potsiou_etal.html (accessed on 18/10/2011). The study aims to demonstrate these technical tools, not only to governmental policy makers, but also to planners, economists, scientists, environmentalists, sociologists and all others with an interest in the life of megacities.

[10] M. Snehmani and P. Mathur (2003). "Restrictions on Supply of Spatial Information on Maps", http://citeseerx.ist.psu.edu/viewdoc/summary?doi=10.1.1.131.4651 (accessed on 18/10/2011).

[11] Ibid.

[12] M. Rao, and M.K.R. Sridhara (2006). "Keeping up with Remote Sensing and GI advances – Policy and Legal Perspectives". *Space Policy* 22, 262-273.

Celestial Bodies of 1967 (the Outer Space Treaty of 1967, OST[13]). Another document that should be mentioned is the Resolution of the United Nations General Assembly from 1986, known as the *Principles Relating to Remote Sensing of the Earth from Outer Space*[14].

The OST, which is the most important legislation for outer space activities, is a treaty adopted by the United Nations in the era when the United States and the USSR were the only space powers in the world. After the launch of 'sputnik', the first satellite launched from earth to outer space in 1957, the United Nations realized that regulations were needed to organize the use of outer space activities, and to ensure that these could be used only for peaceful purposes[15]. The OST prescribes maximum cooperation between the Member States[16], in order to get the maximum benefits from outer space, keeping in mind that any space exploration should only be for the benefit of and in the interest of all countries. This also entails that the dissemination and use of HR data from the satellites in outer space should only be for this objective.

In order to understand the development of the satellite industry, some background history of earth observation data from satellites should be provided. Space-based remote sensing started in 1891 when Germans developed the rocket propelled camera system for the first remote sensing activities worldwide[17]. In 1959, Explorer-6 transmitted the first picture of earth from a satellite, but regular earth observation was really started by TIROS-1[18] in 1960 for meteorological purposes. In the same year, the U.S. started mission CORONA[19], the first operational space imaging reconnaissance satellite programme, using satellite photography for intelligence purposes. This CORONA mission caused a worldwide dispute, when other countries' sovereignty was pierced through remote sensing technology. CORONA's objective was to take pictures from space of the Soviet Bloc countries and use the photographic film for gathering intelligence on the Soviet military capabilities, particularly

13 United Nations (s.d.). *Treaties and Principles on outer space*, http://www.oosa.unvienna.org/pdf/publications/STSPACE11E.pdf (accessed on 26/08/2010).

14 United Nations General Assembly (1986). *Principles relating to remote sensing of the Earth from space*, http://www.un.org/documents/ga/res/41/a41r065.htm (Accessed on 26/08/2010).

15 M. Adhikari (2010). *Legal Perspective of Remote Sensing Data Dissemination for Earth Observation with Special Reference to India*, M.Phil. Dissertation. Hyderabad: NALSAR University of Law. The subject of outer space was taken into consideration by the UN in November 1957, one month after the launch of Sputnik I. In 1958, the UN General Assembly decided to include proposals from the then USSR, and USA into its Agenda under the title 'Questions of Peaceful Uses of Outer Space'. This Assembly created an Ad Hoc Committee in the same resolution. This Ad Hoc Committee decided to further establish a legal sub-committee and technical sub-committee. The first negotiations of the Committee in 1961 led to a unanimous adoption of the UN General Assembly Resolution 1721 (XVI) on 20th December 1961. After that the Legal Subcommittee formulated a set of principles in the form of a draft declaration, which was then adopted by the UN General Assembly on 13 December 1963 as Resolution 1962 (XVIII). This declaration was the basic framework of what would become the 1967 Outer Space Treaty.

16 See the 4th paragraph of the preamble of the OST: "Desiring to contribute to broad international cooperation in the scientific as well as the legal aspects of the exploration and use of outer space for peaceful means."

17 P. Rosenberg (1958). "Earth Satellite Photogrammetry". *Photogrammetry Engineering* 24(3), 353-260.

18 For more details, see http://nasascience.nasa.gov/missions/tiros (accessed on 23/10/2010). The TIROS Programme (Television Infrared Observation Satellite) was NASA's first experimental step to determine if satellites could be useful in the study of the Earth. Its main objective was to test experimental television techniques designed to develop a worldwide meteorological satellite information system.

19 http://space.jpl.nasa.gov/msl/Programs/corona.html (accessed on 23/10/2010).

with regard to nuclear weapons[20]. The 1960s Gemini[21] and Apollo[22] missions were also meant for space photography, this time for scientific purposes, as many scientists had realized the significance of space photography for future earth observation purposes.

The United States in 1972 launched the first satellite for land observation[23], called Earth Resource Technology Satellite (ERTS-1). It can be said that ERTS-1 was the first satellite of its kind in the history of the acquisition of remote sensing data from outer space. In 1975, LANDSAT[24] was launched and, by 1999, IKONOS[25] provided one metre resolution data[26]. While satellite observation was originally developed for military purposes, its civil applications for societal development caught up in the later years[27]. In addition, the initial earth observation strategy of using the space system to monitor land, ocean and atmosphere separately developed into more simultaneous observation activities covering various disciplines to study global scale processes that shape and influence the earth system[28].

The provisions of the Outer Space Treaty

This section discusses the provisions of the OST that may be relevant for finding possible solutions for the conflict between the availability of HR data and the protection of national security.

As a general principle, Article I[29] of the OST prescribes the use of outer space for the benefit and interest of all countries. Such use is in accordance with international law principles on international peace and security, as stated in the preamble[30] to the Charter of

[20] http://spacecovers.com/articles/article_corona2.htm (accessed on 08/07/2011).

[21] http://www-pao.ksc.nasa.gov/history/gemini/gemini.htm (accessed on 22/10/2010). The Gemini Programme (two astronauts to a spacecraft) was a major step forward in U.S. space capability. Intended solely to prepare for Apollo, ten manned missions flew in 1965-66, accomplishing orbital rendezvous, extravehicular activities, tethered flight, and many scientific experiments including 70 mm terrain and weather photography. The Synoptic Terrain Photography Experiment (S005) produced some 1,300 usable 70 mm colour pictures of the Earth's surface. These were an enormous stimulus to space-borne remote sensing, partly because they were released to the public without restriction.

[22] http://www.nasa.gov/mission_pages/apollo/index.html (accessed on 22/10/2010): "Six of the missions – Apollos 11, 12, 14, 15, 16 and 17 – went on to land on the moon, studying land mechanics, meteoroids, seismic, heat flow, lunar ranging, magnetic fields and solar wind".

[23] http://www.eorc.jaxa.jp/en/hatoyama/satellite/satdata/landsat_e.html (accessed on 22/10/2010).

[24] http://landsat.gsfc.nasa.gov/ (accessed 22/10/10). The Landsat Programme is a series of Earth-observing satellite missions jointly managed by NASA and the U.S. Geological Survey. Since 1972, Landsat satellites have collected information about Earth from space. This science, known as remote sensing, has matured with the Landsat Programme. Landsat satellites have taken specialized digital photographs of Earth's continents and surrounding coastal regions for over three decades, enabling people to study many aspects of our planet and to evaluate the dynamic changes caused by both natural processes and human practices.

[25] http://www.absoluteastronomy.com/topics/IKONOS (accessed on 22/10/2010).

[26] M. O'Connell (2009). "Advantage, Geospatial Industry". *GIS Development*, 37.

[27] G. Joseph (2008). *Fundamentals of Remote Sensing. 2nd ed.* Hyderabad: Universities Press.

[28] Ibid.

[29] Article I OST: "The exploration and use of outer space, including the moon and other celestial bodies, shall be carried out for the benefit and in the interests of all countries, irrespective of their degree of economic or scientific development, and shall be the province of all mankind. Outer space, including the moon and other celestial bodies, shall be free for exploration and use by all states without discrimination of any kind, on a basis of equality and in accordance with international law, and there shall be free access to all areas of celestial bodies. There shall be freedom of scientific investigation in outer space, including the moon and other celestial bodies, and states shall facilitate and encourage international co-operation in such investigation".

[30] See the 6th paragraph of the preamble of the UN Charter: "to unite our strength to maintain international peace and security".

the United Nations. Hence, satellite imagery taken from outer space has to be used for the benefit and in the interest of all countries, and in this way should not constitute a threat to national security.

Under Article VI[31] of the OST, the Member States to the Treaty bear international responsibility for any national activity in outer space, whether carried out by governmental agencies or non-governmental entities, such as the dissemination of remote sensing data from any country's satellite. Hence, in the case of any remote sensing activity, even by a non-governmental entity, the corresponding State authority will have to authorize this activity and maintain supervision, because international responsibility under the Treaty will impose on the State authority only. If any international organisation is involved in the outer space activities, then the responsibility will be shared between the organisation and the States involved. For instance, if satellite data were disseminated for prejudicial purposes, e.g. threatening the national security interest of another country, the Member State from the territory or jurisdiction of which the harmful activities are taking place will be liable for these activities under international law. This is also supported by Principle XIV[32] of the Principles Relating to Remote Sensing of the Earth from Outer Space[33].

In addition to the responsibilities under Article VI, Article VII[34] of the OST provides that if a State launches a remote sensing satellite, or a satellite is launched from its territory or facility, then it is internationally liable for any damage to another State or its natural or legal persons caused by the remote sensing satellite. When the OST was adopted, the damage the Member States had in mind related to the crash of a satellite, or part thereof, on to the Earth's surface. But, as all remote sensing activities fall under the OST, including the dissemination of HR satellite data[35], it could be argued that security threats caused by the dissemination of satellite data also qualify as damage under the OST. While the damage is not caused directly by the launching of a remote sensing satellite, the launch has indirectly caused damage to other Member States of the Treaty. However, one could question how far a security threat can be qualified as damage, or whether this security threat needs to have materialized into actual damage to people, institutions or places. This has to be decided by the courts on a case-by-case basis.

---

[31] Article VI OST: "States Parties to the Treaty shall bear international responsibility for national activities in outer space, including the Moon and other celestial bodies, whether such activities are carried on by governmental agencies or by non-governmental entities, and for assuring that national activities are carried out in conformity with the provisions set forth in the present Treaty. The activities of non-governmental entities in outer space, including the Moon and other celestial bodies, shall require authorization and continuing supervision by the appropriate State Party to the Treaty. When activities are carried on in outer space, including the Moon and other celestial bodies, by an international organisation, responsibility for compliance with this Treaty shall be borne both by the international organisation and by the State Parties to the Treaty participating in such organisation".

[32] Principle XIV Principles Relating to Remote Sensing of the Earth from Outer Space: "In compliance with article VI of the Treaty on Principles Governing the Activities of States in the Exploration and Use of Outer Space, including the Moon and Other Celestial Bodies, States operating remote sensing satellites shall bear international responsibility for their activities and [...] irrespective of such activities are carried out by governmental or non-governmental entities or through international organisations to which such States are parties".

[33] Cf. infra.

[34] "Each State Party to the Treaty that launches or procures the launching of an object into outer space, including the Moon and other celestial bodies, and each State Party from whose territory or facility an object is launched, is internationally liable for damage to another State Party to the Treaty or to its natural or juridical persons by such object or its component parts on the Earth, in air space or in outer space, including the Moon and other celestial bodies".

[35] Cf. supra on Principle III of the OST.

Article XI of the OST[36] provides another safeguard from the security threats caused by the dissemination of HR data. This article states that the State parties to the Treaty agree to inform the Secretary General of the United Nations, the public and other international organisations about the details of their activities in outer space, in order to promote international co-operation in the peaceful exploration and use of outer space. So any country that intends to perform remote sensing activities from outer space has to inform the above bodies to the maximum extent possible regarding its activities, including the dissemination of the HR data. After such detailed disclosure of its remote sensing activities, how can a State think of using or disseminating HR satellite data in way that causes security threats to another State or States?

The Principles Relating to Remote Sensing of the Earth from Outer Space

Another legal document that is of importance in the field of remote sensing is the 1986 Principles Relating to Remote Sensing of the Earth from Outer Space[37]. These 15 Principles still remain a non-binding guideline for international remote sensing activities, even though the remote sensing technology has developed strongly since then[38]. In this section, I will discuss the principles that are most relevant with regard to the issue of the potential security threats of HR imagery.

Principle I[39] contains the definition of remote sensing activities. This includes the dissemination of processed data, so the distribution of HR data is explicitly made subject to the Principles. As was indicated earlier, this was not the case in the OST. Therefore, the inclusion of dissemination activities makes it easier to apply the Principles to the question of potential national security threats from HR data than the OST.

Principle II[40] prescribes that remote sensing activities should be carried out for the benefit and in the interest of all countries, irrespective of their degree of economic, social or scientific and technological development, and with particular attention to the needs of developing countries. If one were to apply this to the dissemination of HR data that might cause a security threat for a particular country, one could consider that even though this

[36] Article XI OST: "In order to promote international co-operation in the peaceful exploration and use of outer space, States Parties to the Treaty conducting activities in outer space, including the Moon and other celestial bodies, agree to inform the Secretary-General of the United Nations as well as the public and the international scientific community, to the greatest extent feasible and practicable, of the nature, conduct, locations and results of such activities. On receiving the said information, the Secretary-General of the United Nations should be prepared to disseminate it immediately and effectively".

[37] S.K. Das (2007). *Touching Lives- The Little Known Triumphs of the Indian Space Programme. 1st ed.* New York: Black Classics.

[38] See J.M. Filho (2001). "Why there is no International Convention on Remote Sensing?", http://www.sbda.org.br/revista/Anterior/1718.htm (accessed on 12/07/11): "The shortest and most direct answer to that question is: it is because the US and some of their allies do not want it. And they don't because they understand that the existing principles perfectly meet the needs of this important activity regulation".

[39] Principle I Principles Relating to Remote Sensing of the Earth from Outer Space: "(a)The term 'remote sensing activities' means the operation of remote sensing space systems, primary data collection and storage stations, and activities in processing, interpreting and disseminating the processed data".

[40] Principle II Principles Relating to Remote Sensing of the Earth from Outer Space: "Remote sensing activities shall be carried out for the benefit and in the interests of all countries, irrespective of their degree of economic, social or scientific and technological development, and taking into particular consideration the needs of the developing countries".

dissemination might benefit another country or a group of countries, this would not be permissible. The benefit has to be for all countries, and not just one or more countries.

Principle III states that the remote sensing activities of the Member States shall be governed by international law together with the *Charter of the United Nations*[41], the OST and the rules of the International Telecommunications Union[42]. As the very purpose of the United Nations under Article 1(1) of the UN Charter is to maintain international peace and security, the dissemination of HR data should not cause a threat to the national security of a UN Member State. This is also confirmed by Article 2(4) of the same Charter, which provides that all members of the UN must refrain from any type of threats to others.

Principle IV[43] reiterates the principle of the OST that any outer space activity has to benefit all countries and has to be conducted on the basis of respect for the sovereignty of all States and people over their own wealth and natural resources, and with due regard to the rights and interests of other States and entities under their jurisdiction. In addition, this principle states that any remote sensing activity shall not be conducted in a way which is detrimental to the legitimate rights and interests of the States being observed. Applying this to satellite imagery, this entails that any HR data of the State being watched cannot be used for any detrimental purpose, including threats to the national security of that State or other Member States.

Principle IX[44] is very important with regard to the dissemination of HR data to the public because, in accordance with Article IV of the *Convention on Registration of Objects Launched into Outer Space* and Article XI of the OST, a State must inform the UN Secretary General when conducting any type of remote sensing activity. In addition, it must provide any other relevant information to other States on request, particularly the developing States that are affected by these activities. So the question is whether a State's remote sensing activity causing possible security threats will be permitted by the UN Secretary General, who has the authority to take action against such threats. The answer is always negative as per Article 2(4) of the UN Charter.

---

[41] United Nations (1945). Charter of the United Nations, http://www.un.org/en/documents/charter/index.shtml (accessed on 01/02/2011).

[42] http://www.itu.int/net/about/legal.aspx (accessed on 22/08/2010).

[43] Principle IV Principles Relating to Remote Sensing of the Earth from Outer Space: "Remote sensing activities shall be conducted in accordance with the principles contained in Article I of the Treaty on Principles Governing the Activities of States in the Exploration and Use of Outer Space, including the Moon and Other Celestial Bodies, which, in particular, provides that the exploration and use of outer space shall be carried out for the benefit and in the interests of all countries, irrespective of their degree of economic and scientific development, and stipulates the principle of freedom of exploration and use of outer space on the basis of equality. These activities shall be conducted on the basis of respect for the principle of full and permanent sovereignty of all States and people over their own wealth and natural resources, with due regard to the rights and interests, in accordance with international law, of other States and entities under their jurisdiction. Such activities shall not be conducted in a manner detrimental to the legitimate rights and interest of the sensed State".

[44] Principle IX Principles Relating to Remote Sensing of the Earth from Outer Space: "In accordance with article IV of the Convention on Registration of Objects Launched into Outer space and article XI of the Treaty on Principles Governing the Activities of States in the Exploration and Use of Outer Space, including the Moon and Other Celestial Bodies, a State carrying out a programme of remote sensing shall inform the Secretary-General of the United Nations. It shall, moreover, make available any relevant information to the greatest extent feasible and practicable to any other State, particularly any developing country that is affected by the programme, at its request".

Principle XII[45] provides that the States being watched have non-discriminatory access at reasonable cost to the satellite data relating to their territory, both primary and processed, as soon as these data are produced. So the country on which HR data are collected and disseminated has the full right to observe whether this is done for purposes beneficial or detrimental to the country or any community in the country. If it thinks that there is a security threat, then the country being watched must inform the UN authority concerned with a view to the peaceful settlement of the dispute[46].

The general conclusion from all these principles is that they are directed at creating mutual benefits from remote sensing activities, including the dissemination of HR data, and that they do not allow the use of HR data for any purpose that may cause national security threats to any other Member State. However, the implementation and enforcement of these principles are very difficult, as they are very general and fluid in nature, and are only part of a non-binding resolution passed in the UN General Assembly. Hence, they can only provide limited protection against potential threats to the national security of a State caused by another State or an entity residing in that State.

Effect of space law in India

Legislation and policy on remote sensing data

Although India does not yet have a national policy with regard to the dissemination of remote sensing data, it fully supported the development of the 1986 UN Principles Relating to Remote Sensing of the Earth from Outer Space. The Government applies the 2011 RSDP to the dissemination to all users of India's remote sensing data. Under this policy regulation, all domestic buyers can purchase Indian remote sensing data products from the National Remote Sensing Centre[47] (NRSC). In 2011, this policy changed, allowing access to data with a resolution up to 1 metre (previously, only data was available with a resolution up to 5.8 metres).

However, one could still question the decision to limit the resolution in a time when data with a higher resolution are available from foreign countries' satellites without any limitation[48]. While the availability of remote sensing data is limited in India, data about Indian territory can be purchased via the Internet from competing foreign-based service providers, including data on sensitive locations such as defence installations, the President's House etc. This is not restricted under the RSDP, which allows Indian

---

[45] Principle XII Principles Relating to Remote Sensing of the Earth from Outer Space: "As soon as the primary data and the processed data concerning the territory under its jurisdiction are produced, the sensed State shall have access to them on a non-discriminatory basis and on reasonable cost terms. The sensed State shall also have access to the available analyzed information concerning the territory under its jurisdiction in the possession of any State participating in remote sensing activities on the same basis and terms, taking particularly into account the needs and interests of the developing countries".

[46] Principle XV Principles Relating to Remote Sensing of the Earth from Outer Space: "Any dispute resulting from the application of these principles shall be resolved through the established procedures for the peaceful settlement of disputes".

[47] http://www.nrsc.gov.in (accessed on 18/07/2011).

[48] N. Gopal Raj (2011). "New remote sensing data policy eases restrictions", http://www.thehindu.com/news/national/article2161892.ece (accessed on 20/07/2011): "But much higher resolution images are available from U.S. commercial satellites. The QuickBird's cameras have a resolution of 0.6 metres; the WorldView-1 (0.5 metres); WorldView-2 (0.46 metres); and GeoEye-1 (0.41 metres). As a result, the views of many Indian cities on Google Maps and Google Earth, which use commercially available imagery, can have a resolution of about half a metre".

remote sensing data and data about the Indian territory from satellites owned by foreign countries to be transmitted via the Internet by the service providers of foreign countries, to both Indian service providers and individual customers. For instance, under this policy Google Earth can provide HR images of India's President's residence, of the armed forces headquarters, of sensitive nuclear installations etc. Concerns have been raised over this practice, among others from India's former President Dr. APJ Abdul Kalam[49].

The availability of Google Earth to Indian users can also not be restricted due to the fact that India is a member of the World Trade Organisation (WTO), as is the U.S., where Google Inc. is based. Under the *General Agreement on Trade in Services* (GATS)[50], which is a part of the WTO agreement[51], the so-called Most Favoured Nation Clause[52] states that all the WTO Member States must receive equal trade in services[53] in international trade. Applied to HR imagery, this means that all U.S. companies will get the same opportunity for trade as domestic companies in India for the provision of services of commercial satellite imageries for geographical information systems and vice versa. Thus, to fulfil the condition of *market access* under GATS, India is bound to grant Google access to the Indian market and allows Google Earth services. India has undertaken such a commitment in its schedule of commitments under GATS and has included this obligation to grant market access without reserving any limitations[54].

So, based on international legislation and policy, the Indian government cannot prevent the use of HR data relating to Indian territory stemming from foreign suppliers, even though it may feel that this forms a threat to the national security of the country. However, it is also confronted with a conflict between national security and the dissemination of its own national HR data. This conflict is based on the national 2005 RTI Act, under which the citizens have the right to obtain all information from the government and the public bodies, subject to the provisions and requirements prescribed in the RTI Act.

With regard to HR data, citizens could for instance request access to data from Bhuvan[55], a service parallel to Google Earth run by the Government of India and providing data for the entire Indian territory. However, in accordance with the RSDP mentioned earlier, only

---

[49] The Economic Times (2006). "Mask sensitive Google Earth images", http://articles.economictimes.indiatimes.com/2006-03-10/news/27438828_1_images-installations-website (accessed on 14/03/2011).

[50] World Trade Organisation (1995). General Agreement on Trade in Services, http://www.wto.org/english/docs_e/legal_e/26-gats.doc (accessed on 18/07/2011).

[51] World Trade Organisation (1947). General Agreement on Tariffs and Trade, http://www.wto.org/english/docs_e/legal_e/gatt47_01_e.htm (accessed on 18/07/2011).

[52] Ibid. See also Part II of Article II of General Agreement on Trade in Services: "With respect to any measure covered by this Agreement, each Member shall accord immediately and unconditionally to services and service suppliers of any other Member treatment no less favourable than that it accords to like services and service suppliers of any other country".

[53] Ibid. See also Part I of Article I (2): "2. For the purposes of this Agreement, trade in services is defined as the supply of a service:

(a) from the territory of one Member into the territory of any other Member;
(b) in the territory of one Member to the service consumer of any other Member;
(c) by a service supplier of one Member, through commercial presence in the territory of any other Member;
(d) by a service supplier of one Member, through presence of natural persons of a Member in the territory of any other Member".

[54] See C.A. Feinaeugle (2007). "The UN Principles on Remote Sensing and The GATS: Conflicts or Peaceful Co-existence?". *Proceedings of the 50th International Colloquium of Law of Outer Space.* Washington: American Institute of Aeronautics & Astronautics, http://www.iafastro.net/iac/archive/browse/IAC-07/E6/4/7237/ (accessed on 11/10/2011).

[55] See http://bhuvan.nrsc.gov.in/ (accessed 23/08/10).

data up to 1 metre resolution is made available[56]. Under the RTI Act, a query can be filed why higher resolution data from Bhuvan is not available to the public. It seems strange that the same data are available freely from a foreign service provider, but not from the national service provider, which has collected the data using public money. Can a country like India, which owns excellent space technology, deny its citizens the benefits of this technology and access to the remote sensing data simply because of an old policy made in 2001 that, even though it was changed in 2011, is still not adjusted to the tremendous development of technology in this field? One could wonder how the NRSC would respond to such a query.

At first sight, an answer might be found in the *Official Secrets Act* dating from 1923[57], which provides in its section 3 a prohibition on making available any information which is 'likely to affect the sovereignty and integrity of India, the security of the State or friendly relations with the foreign States'. However, there is no codified definition of secret or official secret, and it is within the discretion of the Government to decide whether it will make this kind of information available. The adoption of the RTI Act changed this. Section 8(2) of the RTI Act clearly states that if the public interest in disclosure outweighs the harm to particular protected interests, such as national security, then access to the requested information cannot be refused. Thus, it seems that the Official Secrets Act cannot be used to deny access. This is confirmed by section 22 of the RTI Act, which explicitly states that the RTI Act supersedes the Official Secrets Act[58]. Thus, if a query is raised about NRSC's high resolution data, the government no longer has the discretion not to reply, but it has to respond with a reason why access is not allowed.

Current socio-political situation in India regarding national security issues

Since the Mumbai terrorist attack in 2008, the Indian Government has become very worried about the security of the country and its people. In this attack, the terrorists used a geographical positioning system (GPS-enabled mobile device) to determine their route to Mumbai. The issue arises how the Government could control such misuse of GPS sets. Comparable questions have been posed with regard to HR satellite data.

Article 21 of the Constitution[59] assures every person the right to life and personal liberty. It is the responsibility of the Government of India to safeguard its law-abiding citizens from terrorist attacks. It wants to do so by restricting access to HR data under the RDSP. However, this access restriction does not seem effective, as terrorist attacks still occur, and these terrorists rely on foreign data to carry out their plans. So what is the point of restricting access to the national HR data? Such questions may come to the Government department concerned as a form of RTI query. Is the Government prepared for that? Some light may be shed on the relationship between the security of India and the availability of HR data to the

56 Cf. infra.

57 See http://www.mha.nic.in/pdfs/Official_Secret_Act1923.pdf (accessed on 26/08/2010).

58 Section 22 RTI Act: "The provisions of this Act shall have effect notwithstanding anything inconsistent therewith contained in the Official Secrets Act, 1923, and any other law for the time being in force or in any instrument having effect by virtue of any law other than this Act."

59 Article 21 Constitution of India, http://lawmin.nic.in/legislative/Art1-242%20(1-88).doc (accessed on 26/08/2010). "No person shall be deprived of his life or personal liberty except according to procedure established by law".

general public by the *Geospatial Data Regulatory Authority Bill*[60], which has been pending since 2009. However, no detailed information on the bill could be obtained so far.

While the government on the one hand wants to restrict the availability of HR data to the public because of national security concerns, on the other hand it also wants to use these data to protect the security of the Indian people. For instance, the government intends to use HR satellite imagery to find the location of the naxalites[61], popularly known as Maoist[62] rebel movements, who are considered to be a threat to the safety of the Indian population[63]. This is not unique: satellite images were also used in the war against Afghanistan (2001) and in the Iraqi war (2003), to find the position of enemies[64]. However, human rights activists argue that the use of satellite imagery to track down the location of the Maoist population groups is violating the right to privacy of the citizen[65].

Interpretation of national security in a new dimension

While this paper has addressed the issue of national security as a political concern where the institutions of the State or the nation are threatened, one might question whether the concept of national security could be interpreted differently in today's society. Besides terrorism and politically inspired crimes, several other issues threaten the security of a country, such as drug cultivation and illicit sale, large scale and systematic piracy, domestic disputes causing people to become refugees, or cyber terrorism. In addition, environmental protection and natural resource conservation have become new security concerns that can seriously impact on our safety and affect the whole world[66]. These new threats change the definition of national security and may be defined under a single term 'human security'[67].

The old traditional conception of national security was important to protect a country from attack by another foreign country or countries. From this perspective, the use of HR data was considered a possible threat to security. However, western countries have shown examples of how HR satellite data are useful for human security[68], so India should follow the same route. Already the ISRO has used satellite imagery to show how Himalayan glaciers

---

[60] Web Newswire (2008). "Cabinet Approval For National Geospatial Data Regulatory Authority Bill Likely By February 2009: Sibal", http://www.webnewswire.com/node/447495 (accessed on 18/09/2011).

[61] See http://en.wikipedia.org/wiki/Naxalite (accessed on 26/08/2010).

[62] S. Aiyar (2010). "Maoism is terrorism", http://indiatoday.intoday.in/story/Maoism+is+terrorism/1/92085.html (accessed on 17/03/2011).

[63] The Hindu (2009). "IAF copters, ISRO images to be used in offensive against Naxals", http://beta.thehindu.com/news/national/article23623.ece?service=mobile (accessed on 26/08/2010).

[64] SpaceTodayOnline (s.d.). "The Satellite Wars", http://www.spacetoday.org/Satellites/YugoWarSats.html (accessed on 17/03/2011): "To prevent the general public from peeking at the war zone from space, the U.S. military began on October 7 paying for the exclusive rights to commercial satellite imagery of Afghanistan even though the nation's spysats were able to take far better pictures. To prevent the commercial firm from selling its pictures of the war zone elsewhere, the U.S. government entered into a multimillion-dollar contract with Space Imaging – a company that sells Earth imagery and information to business, consumer and government customers".

[65] X. (s.d.). "Interview with Susan Wolfinbarger". Geospatial Today. http://geospatialtoday.com/gst/index.php?option=com_content&view=article&id=1789%3Aeyes-in-the-sky&catid=49%3Ainterviews&Itemid=98 (accessed on 17/07/2011).

[66] Y. Hashimoto (2003). "Remote Sensing Satellite and Promotion of Regional Security". *Proceedings of the 46th International Colloquium of Outer Space*. Washington: American Institute of Aeronautics & Astronautics, 234-238.

[67] http://en.wikipedia.org/wiki/Human_security (accessed on 26/08/2010).

[68] Institute for the Protection and Security of the Citizen (2010). "Highlight 2010. Haiti Earthquake: damage assessment in support of relief and reconstruction efforts", http://isferea.jrc.ec.europa.eu/Activities/Technologies/Pages/VeryHighResolutionSatelliteData.aspx (accessed on 26/08/2010).

melt due to climate change[69]. As India has a higher population density than many other countries and faces many problems with regard to human security as well as regarding its overall planning and development, it must insist on the dissemination of HR data to the public rather than limiting it, and go beyond the age old concept of national security, so data can be fully utilized for the wellbeing of the citizens.

Solution in Indian Constitution

When there is any major legal problem in the Indian legislation, the solution is sought in the Constitution of India. Thus, this section examines whether there is a constitutional remedy to support the ability to disseminate Indian HR data to the public.

Article 21[70] of the constitution provides the right to life to every citizen of India. Regarding this right to life, a landmark judgement was passed by the Supreme Court of India in *Maneka Gandhi v. Union of India*[71]. The petitioner Maneka Gandhi's passport was impounded by the Government. This was considered by the Supreme Court an infringement of personal liberty under Article 21(a) of the Constitution because 'due process of law' was not established. One could argue that Article 21 should be extended to the dissemination of HR data for taking care of a State's human security, which is an integral part of the right to life for every citizen of India. The argument is that the citizens will get more accurate information relating to climate change and environmental degradation, personal property land holding, as well as acquisition and distribution of land[72] by the Government itself etc. This would increase their 'human security'. Hence Article 21 broadens the concept of national security of a State, touching the human security issues, and supports the dissemination of HR data for the betterment of the Indian people.

Conclusion

Technology is changing day by day, and it cannot be stopped. There was no national security threat from HR data during the time when it was solely used by militaries or defence personnel in India, but those days are gone. Now the commercialization of satellite technology brings several uses of HR data besides the original military reconnaissance purposes. However, the possible misuse of these data should not automatically lead the Indian government to restrict their availability. When a person commits a crime with a weapon, then it is not that weapon but the very mindset of that person that is responsible for committing such crime. Similarly, it is not the HR data but their use for a particular purpose that brings with it the

---

[69] The Times of India (2010). "Himalayan glaciers shrank 16% in 50 yrs: Isro", http://articles.timesofindia.indiatimes.com/2010-03-28/global-warming/28133759_1_himalayan-glaciers-climate-change-basins (accessed on 18/07/2011).

[70] Article 21 Constitution of India: "Protection of life and personal liberty. No person shall be deprived of his life or personal liberty except according to procedure established by law".

[71] AIR 1978 SC 597. See also http://www.indiankanoon.org/doc/1766147/ (accessed on 19/03/2011).

[72] Daily News and Analysis (2011). "Ex-post facto approval biggest problem in land acquisition: Chief justice", http://www.dnaindia.com/india/report_ex-post-facto-approval-biggest-problem-in-land-acquisition-chief-justice_1569014 (accessed on 28/07/11).

security threat. The Government of India should consider the issue from this point of view rather than just automatically restricting the public availability of HR data.

While India may have a problem with its national security, it should not completely ignore the potential and opportunities of HR data to keep this security intact. Hence the Government should open up the market in HR data, subject to the condition that there must be some specific and strict enactment following the Constitution to punish any infringement on national security.

Relationship between the Public and Private Sector

Part 3

The Relationship between the Public and Private Sector in the Spatial Information Market: How to Define their Roles?

Katleen Janssen[1] *and Niels Vandezande*[2]

Introduction

Traditionally, the public sector has played a large role in the European geographic information market. On the one hand, public sector bodies have historically been the largest collectors of geographic data, often having a monopoly position, due to the enormous efforts and sunk costs required for the collection of such data. These data are re-used by the private sector to create information products and services for their business clients or end users. On the other hand, many public bodies holding geographic data have also started to understand the value of their resources and have been creating and selling value added products in order to gain extra revenues next to their funds obtained from central budget.

Several studies have attempted to estimate the value of public sector geographic data. In 2000, the PIRA study on the commercial exploitation of public sector information (PSI) in the European Union found that the geographic data sector represented more than half of the economic value attributed to PSI, i.e. 'the part of national income attributable to industries and activities built on the exploitation of public sector information'[3]. This economic value was estimated at €68.5 billion, with the geographic data sector accounting for €36 billion of the 68.5. Earlier, a 1996 British study of the economic contribution of Ordnance Survey, the United Kingdom's national mapping agency, determined that the use of Ordnance Survey products and services as a primary input for several key sectors in the economy (e.g. utilities, local government and transport) amounted to a value of £79-£136 billion worth of gross value added[4]. In 2008, the geo-information market in Germany alone was valued at €1.5 billion and the European market for weather and climatologic services based on PSI was estimated around €530 million[5].

The value and potential of geographic data for the creation of services on the information market has led to many discussions between the public sector data holders and the private sector on the role of the former on this information market. This role is twofold: on the one hand, the public sector bodies provide geographic data as a resource to the private sector, while on the other hand they may also be competing with the private sector in the provision of value added information services. Private companies are worried that the public sector

[1] Interdisciplinary Centre for Law and ICT, KU Leuven, katleen.janssen@law.kuleuven.be.
[2] Interdisciplinary Centre for Law and ICT, KU Leuven, niels.vandezande@law.kuleuven.be.
[3] Pira International (2000). *Commercial Exploitation of Europe's Public Sector Information. Final Report*, http://ec.europa.eu/information_society/policy/psi/docs/pdfs/pira_study/commercial_final_report.pdf (accessed on 11/05/2011).
[4] Oxera (1999). *The Economic Contribution of Ordnance Survey GB*, http://www.ordnancesurvey.co.uk/oswebsite/aboutus/reports/oxera/oxera.pdf (accessed on 30/05/2011).
[5] M. Fornefeld et al. (2009). *Assessment of the Re-use of Public Sector Information (PSI) in the Geographical Information, Meteorological Information and Legal Information Sectors*, http://www.epsiplatform.eu/media/files/micus_studie_psi_eu_long (accessed on 20/02/2011).

bodies will abuse their monopoly position on the upstream market of geographic data to influence or even distort the competition on the downstream market of information services[6]. Further, they question whether the public sector should even be active in the downstream market.

This chapter addresses this debate on the role of the public sector on the European market for geographic information services in a number of steps. First, it will explain that this debate, which has also taken place in any market where the public sector has played a role, has been given a new dimension by the introduction of the European Directive on the re-use of PSI in 2003 (PSI Directive)[7]. Next, it will go into some more detail on the possible concerns and barriers raised by the activities of the public sector on the geographic information market on the basis of the principles of European competition law. Finally, it will address some of the criteria used for assessing whether public sector bodies should be part of the information market at all.

The PSI directive: an accelerant to the debate on the role of the public sector

Background

The PSI Directive was adopted in 2003 on the initiative of the European Commission, which felt that the European information industry was not reaching its full potential and that the economic value of public sector data was not being fully exploited. This was at least partially caused by the considerable differences in the rules and practices of the Member States with regard to the re-use of PSI[8]. However, the PSI Directive also wanted to provide a solution for the obstacles created by the behaviour of some public bodies on the market with regard to exclusive deals, refusals to supply data, excessive pricing and unfair competition[9].

The Directive intended to establish a minimum set of rules governing the re-use and the practical means of facilitating re-use of existing documents held by public sector bodies[10], in order to ensure 'fair, proportionate and non-discriminatory conditions for the re-use of such information'[11]. It set a minimum standard for the Member States, but encouraged them to go

[6] Y. Volman (2004). "Exploitation of Public Sector Information in the Context of the eEurope Action Plan". In Aichholzer, G. and Burkert, H (ed.). *Public Sector Information in the Digital Age. Between Markets, Public Management and Citizens' Rights*. Cheltenham: Edward Elgar Publishing Ltd, 96-107.

[7] European Parliament and Council (2003). Directive 2003/98/EC of 17 November 2003 on the re-use of public sector information, *OJ L* 245, 90-96.

[8] See Recital 6 of the PSI Directive.

[9] European Commission (2001). *Communication to the European Parliament, the Council, the European Economic and Social Committee and the Committee of the Regions. eEurope 2002: Creating an EU framework for the exploitation of public sector information*. COM (2001) 607 final.

[10] See Article 1.1 of the PSI Directive.

[11] See Recital 8 of the PSI Directive.

beyond these minimum rules and adopt more open data policies[12]. The Directive had to be transposed by July 2005[13], and was reviewed in 2008 and 2010[14].

Main principles of the PSI Directive

The PSI Directive holds two main types of rules: first, it determines some conditions the public bodies have to follow when making their data available for re-use; and second, it lays down some ground rules for maintaining a level playing field between competitors in the re-use market[15]. The Directive applies to existing documents held by public sector bodies[16], and requires the Member States to 'ensure that, where the re-use of documents held by public sector bodies is allowed, these documents shall be re-usable for commercial or non-commercial purposes in accordance with the conditions' set out in the Directive[17]. Hence, the PSI Directive does not directly impose an obligation on the Member States to allow re-use, but leaves the decision with the Member States or the public sector bodies. If they decide to allow re-use, they have to follow the rules of the Directive. This also entails the obligation to make public sector documents available for both commercial and non-commercial purposes. Hence, it is not possible to make documents available for e.g. research purposes, while refusing commercial re-use.

Conditions for re-use

The PSI Directive requires the public sector bodies to make their data available within a particular time limit[18]. When refusing a request, they have to communicate the grounds of refusal to the applicant, and mention the possible means of redress in case the applicant wishes to appeal the decision[19]. If a request is granted, the public sector documents have to be made available in any pre-existing format or language, through electronic means where possible[20]. The public sector bodies cannot be required to create or adapt any documents in order to comply with a request. In addition, they should only provide abstracts of a document if this does not involve disproportionate effort going beyond a simple operation.

---

[12] M. Fornefeld et al. (2009). *Assessment of the Re-use of Public Sector Information (PSI) in the Geographical Information, Meteorological Information and Legal Information Sectors*, http://www.epsiplatform.eu/media/files/micus_studie_psi_eu_long (accessed on 20/02/2011).

[13] In reality, full transposition was only notified by the last Member State in 2008, and in August 2011 the European Commission is still pursuing a complaint against Poland for incomplete transposition. (see http://ec.europa.eu/information_society/policy/psi/actions_ms/implementation/index_en.htm (accessed on 20/02/2011).

[14] Cf. infra.

[15] For an extensive overview of the provisions of the PSI Directive, see K. Janssen (2010). *The Availability of Spatial and Environmental Data in the European Union. At the Crossroads between Public and Economic Interests.* Alphen a/d Rijn: Kluwer Law International.

[16] Some types of documents are excluded from the field of application: documents the supply of which is an activity falling outside the scope of the public task of the public sector bodies concerned; documents for which third parties hold intellectual property rights; documents which are excluded from access by virtue of the access regimes in the Member States, including on the grounds of the protection of national security, defence or public security, or statistical or commercial confidentiality; documents held by public service broadcasters and their subsidiaries, and by other bodies or their subsidiaries for the fulfilment of a public service broadcasting remit; documents held by educational and research establishments; and documents held by cultural establishments (see Article 1.2 of the PSI Directive).

[17] See Article 3 and Recital 9 of the PSI Directive.

[18] See Article 4.1 of the PSI Directive.

[19] See Article 4.3 of the PSI Directive.

[20] See Article 5.1 of the PSI Directive.

In short, while public sector bodies should make an effort to make their data available, the burden on their time and resources should not be too high as this could discourage them from opening up their data to the re-users.

Another important condition for the public sector bodies is formed by the limits on the possible charges they can impose on the re-use of their documents. The total income from these charges cannot exceed the cost of collection, production, reproduction and dissemination, together with a reasonable return on investment[21]. The charges have to be cost-oriented, but a considerable margin of appreciation is left to the public sector bodies, in order to take into account the possible different funding and business models of the data providers. However, the PSI Directive encourages the public sector bodies to charge for their data at the level of the marginal costs for reproduction and dissemination[22]. Whatever pricing model is used, the standard charges have to be pre-established and published, through electronic means where possible and appropriate, together with any applicable conditions[23]. In addition, standard licences have to be made available by the Member States that can be adapted to meet particular licence applications, are available in digital format and can be processed electronically. The Member States should encourage the public sector bodies to use these standard licences when making their data available for re-use[24]. Finally, the Member States should also take some measures to ensure that potential re-users are not just aware of the conditions and charges for the re-use of public sector documents, but that they also can find the documents and know where to request them. Such practical arrangements to facilitate the search for re-usable documents could include the creation of asset lists or portal sites[25].

Level playing field

One of the main concerns of the PSI Directive was to ensure that an open information market could develop, enabling new players to enter the market and the private sector to compete with the public bodies on an equal level. Many public sector data providers had a history of giving preferential treatment to particular private partners (either by granting exclusive rights or charging lower prices for the use of their data) or unfairly competing by charging very low prices for their value-added products, often combined with high charges for the underlying data. The PSI Directive provides some sector-specific regulations to limit such behaviour.

The PSI Directive requires any conditions on the re-use of public sector documents to 'not unnecessarily restrict possibilities for re-use and [to] not be used to restrict competition'[26]. Unfortunately, no additional guidance is given on the criteria to determine when particular conditions would be restricting competition. The directive only adds that any conditions

---

[21] See Article 6 of the PSI Directive.

[22] See Recital 14 of the PSI Directive.

[23] See Article 7 of the PSI Directive. In case of atypical cases for which the standard charges do not apply, the public sector bodies have to indicate the factors that are taken into account in the calculation of the charges.

[24] However, the public sector bodies are under no obligation to use these standard licences, so they can keep using their own licensing models. Many initiatives have been taken to harmonise licensing conditions and templates, either in a particular country or in a particular sector (e.g. the UK's Open Government Licence, the Queensland Government Information Licensing Framework, and the Dutch Geo-Shared initiative, which are discussed in the chapter of Van Loenen, Janssen and Welle Donker).

[25] See Article 9 of the PSI Directive.

[26] See Article 8.1 of the PSI Directive.

for re-use have to be non-discriminatory for comparable categories of re-use[27]. Again, this should ensure a level playing field for the market players, but the same problem arises: how to assess when there might be discriminatory conditions? More insight in the concept of 'comparable categories' of re-use is needed. At first sight, it would seem clear that commercial and non-commercial use can be seen as non-comparable categories, and within these categories, different conditions might be suitable for e.g. value adding services and pure reproduction of the data. However, even commercial and non-commercial use might not be so easy to distinguish. What is the difference between some 'techies' developing a new application in their basement for fun, and the development of a similar application by a commercial company (which may be based on earlier work of other 'techies')? In short, the comparable character of types of re-use is difficult to determine.

One specific category of re-use is targeted separately by the PSI Directive, i.e. the re-use of public sector documents by a public sector body as input for its own commercial activities that fall outside the scope of its public tasks[28]. In this case, the public sector bodies have to apply the same conditions and charges to their own re-use as they do to other users. This should reduce the 'risk that the public sector body involved could abuse its privileged position to gain a competitive advantage over other market players'[29]. In order to fully implement this requirement, the public sector bodies most likely will need to maintain separate accounts or move their commercial activities to a body with a separate legal personality.

Finally, the PSI Directive also ensures that the public sector bodies can no longer provide advantages to one or more particular private partners by giving them an exclusive right to re-use their data. Re-use has to be open to all potential actors in the market, and contracts or other arrangements between public sector bodies and third parties should not grant exclusive rights[30]. Exceptions to this can only be made in case an exclusive right would be necessary for the provision of a service in the public interest[31]. An example of such a situation is if no commercial publisher would make the investment to publish the information without an exclusive right.

The review of the PSI Directive

In 2008, the European Commission organised a review of the application and impact of the PSI Directive, and it launched a consultation with the Member States and the stakeholders in the European Union. While the impact of the PSI Directive was found positive, concerns were raised that important barriers to the European re-use market still remained. These barriers included, next to a lack of transparency and strict licensing conditions, competition between the private and the public sector. Questions arose about the limits of the public

[27] See Article 10.1 of the PSI Directive.
[28] See Article 10.2 of the PSI Directive.
[29] European Commission (2002). *Proposal for a Directive of the European Parliament and of the Council on the re-use and commercial exploitation of public sector documents*. COM (2002) 207 final.
[30] See Article 11.1 of the PSI Directive.
[31] See Article 11.2 of the PSI Directive.

task when public sector bodies commercially competed with the private sector, and about unfair competition practices of public sector bodies[32].

As the results of the review were considered inconclusive because the national transpositions of the PSI Directive were too recent to show much impact, the European Commission decided to organise a new review of the directive in 2010. This review was focused on the scope of application of the PSI Directive, the non-mandatory character of re-use, and the charging provisions. However, other comments were also welcomed, and again the private sector protested the anti-competitive behaviour of the public sector bodies[33].

At the time of writing of this chapter, the European Commission has not proposed any amendments to the Directive yet. However, it is expected that the Commission will aim to introduce a general obligation for the Member States and the public sector bodies to allow re-use of their generally accessible documents.

Public sector activities on the geographic information market

Public bodies traditionally have a strong position on the market for geographic data and services, either by supplying geographic data to the private sector for their commercial applications, or by creating added-value information services themselves and selling these on the market. This strong position stems from the public sector's natural monopoly in the collection and processing of geographic data. However, the development of technology and the information society has increasingly enabled the private sector to create and use geographic data in its products and services. As the collection of geographic data, particularly reference data, is still very expensive and time-consuming and the public sector still has a legal monopoly on some types of data, the private companies still mostly look towards the public geographic data holders for obtaining these data. This had led to a situation where the private sector is both the customer and the competitor of the public sector geographic data holder.

Public bodies may be tempted to use this situation to maximise their short term revenues without taking into account the need for a level playing field, to increase their revenue by setting higher prices for the data, or to extend their monopolies to the downstream market[34]. In addition, cross-subsidisation may occur if the public body uses funds from its public service activities to 'sponsor' its commercial activities[35]. Such activities may lead to a distortion of the market and unfair competitive behaviour of the public sector. As was mentioned earlier, the PSI Directive has tried to limit this behaviour by imposing some rules

---

[32] European Commission (2009). *Communication to the European Parliament, the Council, the European Economic and Social Committee and the Committee of the Regions. Re-use of Public Sector Information – Review of directive 2003/98/EC.* COM (2009) 212 final.

[33] European Commission (2011). *Results of the online consultation of stakeholders. Review of the PSI Directive,* http://ec.europa.eu/information_society/policy/psi/docs/consultations/cons2010/Results%20of%20the%20online%20consultation%20of%20stakeholders%20final.doc (accessed on 15/04/2011).

[34] Y. Volman (2004). "Exploitation of Public Sector Information in the Context of the eEurope Action Plan". In Aichholzer, G. and Burkert, H (ed.). *Public Sector Information in the Digital Age. Between Markets, Public Management and Citizens' Rights.* Cheltenham: Edward Elgar Publishing Ltd, 96-107; H. Perritt Jr. (1995). "Should local governments sell local spatial databases through state monopolies?". *Jurimetrics Journal* 35, 449-469.

[35] J. Peters and R. Vriesendorp (2002). "Insolvency of public entities other than the State under Dutch law". *Electronic Journal of Comparative Law,* http://www.ejcl.org/64/art64-23.html (accessed on 15/04/2011).

on the provision of data held by public sector bodies, but complaints about the attitude of the public sector still remain.

These complaints should be divided into two issues: on the one hand, there is the issue of fair competition and how the public bodies have to act on the market if they offer commercial services; on the other hand, there is also the question whether, even if they follow the rules of the level playing field, the public bodies should even provide the service, either as a commercial activity or as a part of their public task. While the first question can be addressed by competition law, the second question is a matter of policy and political decisions. It is no longer about the procedural matter whether the public bodies are obeying the rules of the market, but rather about the policy decision whether that market should be entered into in the first place or whether the public body should stay away from offering an information service and leave this to the private sector. In addition, the public body is confronted with the question whether a particular service it wants to offer is actually a market service, or if it is part of its public services that remain outside of any market environment.

In the following sections, we first address the conditions for public sector bodies providing services on the market and the possible infringements these bodies can make on the competition provisions ruling the market. Then we move to the policy question whether the public sector should enter the market and provide information services or whether it should even consider offering particular information services as part of its public service activities.

The provision of information services by public bodies under EU competition law

In the European Union, if public bodies act on the market in competition with the private sector, they have to follow the rules of European competition law laid down in Article 101 and following of the Treaty on the Functioning of the European Union (TFEU). These provisions require undertakings to refrain from anticompetitive behaviour on the market, based on the idea that a level playing field on the market ensures the most equitable conditions for all players on the market and is the most advantageous for the consumers.

Public bodies as undertakings

If public bodies act as undertakings on the market, they are subject to the rules of Article 101 and following TFEU. Undertakings have been defined by the Court of Justice of the European Union (ECJ) as 'every entity engaged in an economic activity, regardless of the legal status of the entity or the way in which it is financed'[36]. Whether charges are imposed or profits are made is irrelevant. Hence, even free of charge services can be considered

[36] ECJ, Case C-41/90, Höfner & Elser v. Macotron GmbH, 23 April 1991, paragraph 21, *European Court Reports* 1991, I-1979.

economic activities[37]. It is sufficient that the activities can be carried out by a private entity, even though this may not actually be the case[38].

A distinction must be made between the acts of Member States or public bodies that are part of the performance of their sovereign or administrative functions and their other activities. The former cannot be considered as economic activities and are not subject to competition law, as they are part of the State's official authority, i.e. 'a task in the public interest which forms part of the essential functions of the State [and] is connected by its nature, its aim and the rules to which it is subject with the exercise of powers ...which are typically those of a public authority'[39]. Examples of such activities include air space management services, pollution control in a harbour, tax collection, and compulsory social security schemes[40]. With regard to social services (e.g. social security, pension funds), another criterion is also taken into account: if social solidarity prevails over the economic nature of the activity, the performing entity is not an undertaking[41].

On the contrary, the ECJ qualified the following activities as economic: state-run employment agencies, independent customs agents, medical aid organisations providing ambulance services, and public television broadcasting organisations[42]. If we apply the same reasoning to geographic information services, there are definitely some services that can be qualified as economic activities under the rules of EU competition law. For instance, many activities of the Dutch Meteorological Office (KNMI) could be considered as economic activities, as either they are already offered by the private sector or they could be, but are currently part of KNMI's monopoly. Such services include general weather reports, expectations for road traffic, historical weather information, expectations for oil rigs or harbours[43]. Hence, these activities are economic activities under EU competition law, and have to follow the rules of a level playing field. Another example is the planned provision of the national road database by the Dutch Executive Agency for Transport, Public Works and Water Management (Rijkswatersstaat)[44]. The Agency decided to make this database openly available on the Internet for any use. As such databases were already created and offered by the private sector, this activity would give the Agency the status of undertaking subject to competition law. While the creation of the database might be considered as part of public policy and in this way be considered an act of official authority, the dissemination of the

---

[37] ECJ, Case C-67/96, Albany International B.V. v. Stichting Bedrijfspensioenfonds Textielindustrie, 21 September 1999, paragraph 85, *European Court Reports* 1999, I-5751.

[38] A. Jones and B. Sufrin (2008). *EC Competition Law: text, cases and materials. 3rd ed.* Oxford: Oxford University Press.

[39] ECJ, Case C-343/95, Cali & Figli Srl v. Servizi ecologico porto di Genova SpA, 18 March 1997, paragraph 23, *European Court Reports* 1997, I-1547.

[40] ECJ, Case C-364/92, SAT Fluggesellschaft mbH v. Eurocontrol, 19 January 1994, *European Court Reports* 1994, I-43; Case C-343/95, Cali & Figli Srl v. Servizi ecologico porto di Genova SpA, 18 March 1997, *European Court Reports* 1997, I-1547; Case C-207/01, Altair Chimica SpA v ENEL Distribuzione SpA, 11 September 2003, *European Court Reports* 2003, I-8875; Case C-159/91 and C-160/91, Poucet et Pistre v. Assurances Générales de France and Caisse Mutuelle Régionale du Languedoc-Roussillon, 17 February 1993, *European Court Reports* 1993, I-637. See also R. Whish (2009). *Competition law. 6th ed.* Oxford: Oxford University Press.

[41] W. Sauter (2008). "Services of general economic interest and universal service in EU law". *European Law Review* 33(2), 167-193; T. Prosser (2005). *The limits of competition law. Markets and Public Services.* Oxford: Oxford University Press.

[42] ECJ, Case C-41/90, Höfner & Elser v. Macotron GmbH, 23 April 1991, *European Court Reports* 1991, I-1979; Case C-35/96, Commission v. Italian Republic, 18 June 1998, *European Court Reports* 1998, I-3851; Case C-475/99, Ambulanz Glöckner v. Landkreis Südwestpflatz, 25 October 2001, *European Court Reports* 2001, I-8089; ECJ, Case 155/73, Sacchi, 30 April 1974, *European Court Reports* 1974, 409.

[43] KPMG (2007). *Ministerie van Waterstaat. Evaluatie van de wet op het KNMI*, http://www.knmi.nl/cms/content/8524/wet_op_het_knmi_geevalueerd (accessed on 16/04/2011) [in Dutch].

[44] See http://computertotaal.nl/nieuws/3701-vrijgave-wegenbestand-stap-dichterbij.html [in Dutch] (accessed on 16/04/2011).

database to the public is an economic activity that can also be performed by the private sector[45].

Possible distortions of competition law by public bodies providing information services

There are several ways in which the EU Member States or their public bodies can distort competition. These can be divided in two main categories. First, the Member States and particularly the public bodies can infringe the competition rules in their acts as undertakings. Second, they may also take decisions as a sovereign power that lead to the preferential treatment of particular private companies. In this section, we will not give a full overview of all the possible manners in which EU competition law could be infringed, but provide some examples on how competition law might deal with the provision of spatial data or geographic information services by public bodies on the market[46].

Agreements between undertakings

Article 101 TFEU (former Article 81 EC-Treaty) forbids agreements between undertakings, decisions by associations of undertakings and concerted practice that may affect trade between Member States[47] that have as their object or effect the prevention, restriction or distortion of competition within the common market. Examples of such behaviour include the fixing of selling prices or any other trading conditions, or the application of dissimilar conditions to transactions with parties outside the agreements.

In the geographic data sector, a group of undertakings of which the market conformity has been questioned is ECOMET, an association of meteorological offices set up to ensure the exchange of meteorological data between the national meteorological services and to organise the availability of weather data to outside parties[48]. The members of ECOMET have agreed on a joint data licensing policy, including standard licences for different user groups, volume discounts and special tariff schemes. However, the exact prices for the use of the data are still determined by the members. The European Commission examined whether there was a possible infringement of Article 101 TFEU, but found there was still adequate competition between the partners and that the organisation worked in such manner that fair competition with independent service providers was still possible[49].

Abuse of dominant position

While infringements of Article 101 TFEU do not occur very frequently, abuse of dominance by undertakings is more common. Under Article 102 TFEU, an undertaking is prohibited

[45] Cf. infra on the possible breach of competition law by Rijkswaterstaat. See also K. Janssen (2010). *The Availability of Spatial and Environmental Data in the European Union. At the Crossroads between Public and Economic Interests.* Alphen a/d Rijn: Kluwer Law International.

[46] For a full overview of the provisions of competition law, see A. Jones and B. Sufrin (2008). *EC Competition Law: text, cases and materials. 3rd ed.* Oxford: Oxford University Press; P. Roth and V. Rose (2008). *Bellamy & Child European Community Law of Competition. 6th ed.* Oxford: Oxford University Press.

[47] This rules out market distortions that only have effect on a smaller scale on the national market.

[48] See http://ecomet.eu/index.php (accessed on 17/04/2011).

[49] M. Kamps (2004). "Thunder and Lightning: Public Sector Information Policy Experiences of Private Meteorological Service Providers". In Aichholzer, G. and Burkert, H (ed.). *Public Sector Information in the Digital Age. Between Markets, Public Management and Citizens' Rights.* Cheltenham: Edward Elgar Publishing Ltd, 163-186.

from abusing its dominant position within a common market or in a substantial part of it, insofar as it may affect trade between Member States. A number of elements have to be present for this prohibition to apply[50]. First, the undertaking has to have a dominant position, i.e. 'a position of economic strength which enables it to hinder the maintenance of effective competition on the relevant market by allowing it to behave to an appreciable extent independently of its competitors and customers and ultimately of consumers'[51]. To determine whether a particular undertaking is dominant, the European Commission and the ECJ rely on the market share of the undertaking, but also on other factors, such as economies of scale and scope, privileged access to supply, the established position of the undertaking on the market, and other strategic barriers to entry for other undertakings[52]. Of course, when an undertaking has a monopoly, the dominance is clear. In other situations, a case-by-case assessment is made. Second, the undertaking has to abuse its dominant position. The concept of abuse was defined by the ECJ in the following manner:

> "The concept of abuse is an objective concept relating to the behaviour of an undertaking in a dominant position which is such as to influence the structure of a market where, as a result of the very presence of the undertaking in question, the degree of competition is weakened and which, through recourse to methods different from those which condition normal competition in products or services on the basis of the transactions of commercial operators, has the effect of hindering the maintenance of the degree of competition still existing in the market or the growth of that competition"[53].

Undertakings in a dominant position have a 'special responsibility' on the market, entailing that behaviour that might be tolerated from other players may be considered to illegally distort competition if it is performed by a dominant undertaking[54]. Abusive acts mainly relate to pricing (either excessive or predatory), or refusals to supply.

Abusive acts with regard to pricing

Price discrimination may be considered an abuse when the same good or service is provided at different prices to different customers in the same relation to the provider, for example when a data provider sells data to one customer cheaper than to another, while both customers are competing manufacturers on the same market who need the product as input for their services[55]. Next, excessive pricing is an obvious way of exploiting a dominant position on the market. A more problematic issue is predatory pricing. This occurs when

---

[50] The actual or potential effect on trade between the Member States will not be discussed here. For more information, see P. Roth and V. Rose (2008). *Bellamy & Child European Community Law of Competition. 6th ed.* Oxford: Oxford University Press.

[51] ECJ, Case 85/76, Hoffman-La Roche & Co. AG v. Commission, 13 February 1979, paragraph 38, *European Court Reports* 1979, 461.

[52] R. Whish (2009). *Competition law. 6th ed.* Oxford: Oxford University Press; A. Jones and B. Sufrin (2008). *EC Competition Law: text, cases and materials. 3rd ed.* Oxford: Oxford University Press.

[53] ECJ, Case 85/76, Hoffman-La Roche & Co. AG v. Commission, 13 February 1979, paragraph 91, *European Court Reports* 1979, 461.

[54] ECJ, Case 322/81, NV Nederlandsche Banden Industrie Michelin v. Commission, 9 November 1983, paragraph 57, *European Court Reports* 1983, 3461.

[55] A. Jones and B. Sufrin (2008). *EC Competition Law: text, cases and materials. 3rd ed.* Oxford: Oxford University Press; P. Roth and V. Rose (2008). *Bellamy & Child European Community Law of Competition. 6th ed.* Oxford: Oxford University Press.

an undertaking maintains such low prices that its competitor cannot follow and is driven from the market. In principle three elements have to be present: (1) a temporary, substantial reduction in price by the dominant firm in response to an anticipated or actual market entry or growth in sales by a smaller competitor; (2) price levels that are unprofitable or barely profitable; and (3) pricing 'aimed at' eliminating or disciplining a specific competitor. Once the competitor is eliminated, the undertaking can raise its prices again and recoup its losses[56]. A comparable abuse is margin squeezing. If an undertaking is active both in the upstream market (e.g. by selling spatial data) and the downstream market (e.g. by providing geographic information services in competition with the customers of the spatial data), it can squeeze out its competitors in the downstream market in two ways. Either it prices its products in the downstream market very low, so that the competitor cannot follow without incurring losses, or it can charge high prices to the competitor for the raw materials in the upstream market[57].

Many examples can be found in the geographic data sector where the question of possible abuse of dominance by public bodies has been raised by the private sector. However, most of these cases have never been brought before a court or competition authority, so it is difficult to assess whether an actual infringement of Article 102 TFEU (or of the equivalent provisions in national competition law) took place.

A possible example of predatory pricing or price dumping might be found in the dissemination of the national road database by the Dutch Executive Agency for Transport, Public Works and Water Management. Making the database freely available on the market, while competitors are required to charge considerable prices to recuperate their costs, could be considered equivalent to price dumping[58]. However, provided that the Agency could even be seen as a dominant undertaking, the free of charge dissemination by the Agency is not aimed at driving the private companies out of the market, so most likely the conditions for predatory pricing will not be fulfilled. However, the impact on the market cannot be denied, so the Agency decided in 2006 to not make the data available for commercial use yet and in summer 2011, it still has not done so.

Another example that might qualify as an abuse of dominance is the price discrimination or possible margin squeezing in the United Kingdom's marine information sector. The Marine Information Council Working Group on Data Access found in 2003 that some of the organisations that held publicly funded marine data competed with private companies on the market, and had free access to their own data while the competitors had to pay[59]. The association of meteorological services ECOMET was also accused by the private sector of abusing its dominance. The European Commission provided the association with a comfort letter in 1999, stating that there was no competition problem and that the association could

[56] P. Roth and V. Rose (2008). *Bellamy & Child European Community Law of Competition. 6th ed.* Oxford: Oxford University Press.

[57] A. Jones and B. Sufrin (2008). *EC Competition Law: text, cases and materials. 3rd ed.* Oxford: Oxford University Press.

[58] M. Van Eechoud and B. Van Der Wal (2008). *Creative commons licensing for public sector information. Opportunities and pitfalls*, http://www.ivir.nl/creativecommons/en_government_information.html (accessed on 17/04/2011); B. Van Loenen et al. (2006). *Beschikbaar stellen van geo-informatie bij Rijkswaterstaat. Analyse van de (on)mogelijkheden van het op korte termijn vrij beschikbaar stellen van vier geo-datasets*, http://www.bastiaanvanloenen.nl/pubs/2006_RWS_AGI_Eindrapport_TUD.pdf [in Dutch] (accessed on 17/04/2011).

[59] United Kingdom Marine Information Council Working Group on Data Access (2003). *Achieving optimal value from publicly funded marine information resources*, http://epsiplatform.eu/media/files/ukmic_achieving_optimal_value (accessed on 17/04/2011).

maintain its activities[60]. However, ECOMET's private sector competitors still complained about discriminatory pricing and margin squeeze problems, because allegedly only some of the association's members followed the principles of equal treatment of independent service providers[61].

Refusal to supply

In some cases, the refusal of a dominant undertaking to supply goods or services can be considered an abuse of dominant position[62]. This also includes making the supply conditional on control over the further processing or marketing, supplying only on discriminatory and unfair conditions, or making an offer that the supplier knows is unacceptable for the customer[63].

An example relevant for the geographic data sector is the 1995 Magill case, which deals with another type of public sector data, i.e. TV listings of a public sector broadcaster. In this case, the ECJ found an abuse of dominance by the Irish national broadcasting organisation (and other broadcasters), because it refused to licence its intellectual property rights (IPR) on the listings of television programmes to the publisher of a commercial magazine[64]. This refusal was considered abusive because of four main reasons: there was no substitute for the listings of the dominant broadcaster; the refusal to licence the IPR prevented the appearance of a new product for which there was potential customer demand; there was no justification for the refusal; and the dominant undertaking was excluding all competition on the secondary market of weekly listings and reserving the market to itself[65].

With regard to geographic data, some examples of refusals to supply can be found that have triggered questions with regard to the abusive character of the refusal. For instance, the United Kingdom's Office of Fair Trading (OFT) found in its 2006 report on the Commercial Use of Public Information that Ordnance Survey in some cases only provided limited access to its unrefined information, while concentrating on developing the refined products itself, so that businesses had fewer opportunities to create competing added-value products[66]. While there was no formal case for abuse of dominance, the OFT did suggest Ordnance Survey to adopt a more flexible licensing policy[67].

A refusal to supply that led to a court case in the early 1990s was the non-delivery of aeronautic data by the French meteorological office, which had a monopoly on the upstream market of the data and a dominant position on the downstream market of added-value information products. The French competition authority found no abuse of dominance,

---

[60] M. Kamps (2004). "Thunder and Lightning: Public Sector Information Policy Experiences of Private Meteorological Service Providers". In Aichholzer, G. and Burkert, H (ed.). *Public Sector Information in the Digital Age. Between Markets, Public Management and Citizens' Rights.* Cheltenham: Edward Elgar Publishing Ltd, 163-186. See also http://www.primet.org/index.php?Itemid=29&id=81&option=com_content&task=view (accessed on 17/04/2011).

[61] Ibid.

[62] CFI, Case T-41/96, Bayer AG v. Commission, 26 October 2000, *European Court Reports* 2000, II-3383.

[63] P. Roth and V. Rose (2008). *Bellamy & Child European Community Law of Competition. 6th ed.* Oxford: Oxford University Press.

[64] ECJ, Case C-241/91P and C-242/91P, Radio Telefis Eireann v. Commission, 6 April 1995, *European Court Reports* 1995, I-743.

[65] J. Kabel et al. (2001). *Kennisinstellingen en informatiebeleid. Lusten en lasten van de publieke taak.* Amsterdam: Universiteit van Amsterdam [in Dutch].

[66] Office of Fair Trading (2006). *The commercial use of public information (CUPI),* http://www.oft.gov.uk/OFTwork/publications/publication-categories/reports/consumer-protection/oft861 (accessed on 17/04/2011).

[67] In the meantime, Ordnance Survey has revised its licensing policy and has adopted a more open data policy, but it is not clear in how far this has changed the relationship with its competitors on the market.

because the office's arguments that the dissemination would harm security and would be a risk for aerial traffic were deemed justified. While the Court of Appeal disagreed with the competition authority and found an abuse, the *Cour de Cassation* did not even consider the provision of aeronautic data to be an economic activity, so competition law was not even applicable. This decision was severely criticised by French scholars[68]. In a later case, the French competition authority had to judge whether the Marine Hydrographical and Oceanographic Service of the Ministry of Defence (SHOM) abused its dominance by offering value added nautical information services for leisure boating in competition with the private sector, while at the same time limiting access of its competitors to the underlying data[69]. The competition authority acknowledged the economic character of the activities and stated that SHOM abused its position by using its IPR to refuse the provision of information to its competitors.

State actions infringing competition law

The EU Member States and the public bodies may not only infringe competition law by their economic activities, but also by their actions as the State or part thereof. First, Article 106(1) TFEU prohibits Member States to enact or maintain in force any measures regarding public undertakings and undertakings granted with special or exclusive rights contrary to the Treaty, particularly with regard to competition or discrimination on the grounds of nationality. So if the Member States have taken measures that lead to the undertaking abusing its dominant position, e.g. by granting an exclusive right on the upstream market that makes the undertaking treat the competitors on the downstream market unfairly, the Member States are also liable for infringing Article 106 TFEU.

A special situation is created for undertakings (whether private or public) that are given the task by the State of providing an economic activity that is in the public interest, e.g. the public supply of utility services such as energy, telecommunication, transport, water, waste disposal or postal services, or the provision of ambulance services, broadcasting services or pension funds. Undertakings that have been entrusted by the Member States with the provision of such services of general economic interest (or services having the character of a revenue-reproducing monopoly[70]) can be exempted from the application of EU competition law, insofar as the application of the rules obstructs the performance, in law or in fact, of the particular tasks that are assigned to them. Hence, if these undertakings cannot provide their services under economically acceptable or balanced conditions, they can for example take refuge to measures of cross-subsidisation or State aid[71]. However, if this is not necessary for the performance of such an economic activity performed in the public interest[72], competition law may be infringed by the undertaking providing the service and by the Member State.

---

[68] J.M. Bruguière (2002). *Les données publiques et le droit.* Paris: Editions Litec [in French]; B. Nouel (1996). "Commercialisation des données publiques et concurrence". In European Commission (ed.). *Proceedings of the European Commission Conference: Access to Public Information. A Key to Commercial Growth and Electronic Democracy*. Stockholm: European Commission, 1-9 [in French]; L. Teresi (2005). "Données publiques et concurrence". *Concurrences* 4, 143-156 [in French].

[69] J.M. Bruguière (2002). *Les données publiques et le droit.* Paris: Editions Litec [in French]; L. Teresi (2005). "Données publiques et concurrence". *Concurrences* 4, 143-156 [in French].

[70] This type of services has received very little attention in case law and literature, and is not relevant for the market of spatial information services. Therefore, they are further disregarded.

[71] K. Janssen (2010). *The Availability of Spatial and Environmental Data in the European Union. At the Crossroads between Public and Economic Interests.* Alphen a/d Rijn: Kluwer Law International.

[72] Cf. infra.

As we have mentioned before, the provision of geographic information services may often be considered as an economic activity, but these services may also be in the public interest and therefore qualify as services of general economic interest. For instance, the provision of the national road database by the Dutch Executive Agency for Transport, Public Works and Water Management could be qualified as an economic activity[73], but the Minister for Transport also stated that the dataset was necessary for policy makers and road management to perform their tasks and that there was a public interest in making the data available due to INSPIRE and the national initiative on authentic sources[74]. In this way, making the road database available could be seen as a service of general economic interest.

Qualification of their geographic information services as a service of general economic interest would give the public bodies some leeway on complying with competition law and give them some space to cross-subsidize some of their economic activities with public money, or to put in place particular conditions on the use of their services. However, it seems unlikely that a public body would be able to argue having a problem with offering a service under economically acceptable conditions in the situations where it is being challenged for predatory pricing or charging dumping prices[75].

Finally, next to taking any measures that may unfairly advantage undertakings with special or exclusive rights of performing services of general economic interest, the State can also distort the market by giving illegal State aid to one or more undertakings in order to subsidize their activities. Under Article 107 TFEU, any aid granted by a Member State or through State resources in any form whatsoever which distorts or threatens to distort competition by favouring certain undertakings or the production of certain goods is incompatible with the common market, insofar as it affects trade between Member States. Examples of State aid are investment grants, loans at reduced interest rates, or preferential fiscal treatment. To determine whether a particular grant or investment from the State could be considered as State aid, it should be examined whether the terms under which the funds are provided go beyond what would be acceptable for a private investor operating under normal market conditions, having regard to the available information and foreseeable developments at that time[76].

While no official complaints about State aid to the European competition authorities have been made with regard to geographic data, the question has been raised in some occasions, for instance with regard to the provision of government funds in the United Kingdom to Ordnance Survey under the National Interest Mapping Services Agreement in the early 2000s. It was questioned by LOCUS, an association of re-users of PSI, whether government funds granted to Ordnance Survey without a public procurement procedure could be considered State aid, as many of the activities that were being funded could be

[73] Cf. supra.

[74] Minister of Transport and Water Management (2007). *Reply to questions from Members of Parliament Roemer and Roefs,* question no. 2060707530, https://zoek.officielebekendmakingen.nl/ (accessed on 17/04/2011) [in Dutch]; Commissie Tweede Consultatieronde Vrijgeven NWB Bestand (2006). *Advies,* http://www.verkeerenwaterstaat.nl/ Images/ Advies%20Cie%20tweede%20consultatie_tcm195-189308.pdf (accessed on 22/05/2009) [in Dutch].

[75] K. Janssen (2010). *The Availability of Spatial and Environmental Data in the European Union. At the Crossroads between Public and Economic Interests.* Alphen a/d Rijn: Kluwer Law International.

[76] P. Roth and V. Rose (2008). *Bellamy & Child European Community Law of Competition. 6th ed.* Oxford: Oxford University Press.

considered economic and subject to competition rules[77]. Follow-up agreements between the government and Ordnance Survey were only concluded after a public procurement procedure, in order to take away any concerns of the private sector[78].

Role of the public sector

In the previous section, we described the possible problems that might occur if public bodies provide geographic information services on the market, in competition with the private sector. In summary, when these public bodies perform economic activities, they have to make sure that they comply with competition law, in order not to distort the market and create unfair disadvantages for their private competitors. However, this does not answer the question whether the public sector should provide information services in the first place. There are two elements to this: first, whether providing a geographic information service that could also be provided by the private sector should also be part of the public body's remit; and second, if it is not, whether the public body should refrain from providing the service altogether. While this is related to the issue of competition law discussed earlier, it is not the same: even if the geographic information service is an economic activity, this does not mean that it cannot or should not be part of the public body's public task or remit.

Several criteria could be considered to determine whether a public body should invest in a particular geographic information service. A first factor is whether the information service in question is part of the public body's tasks as they are determined by the law governing its formation, administration and activities, by its own bylaws, policy documents or common practice[79]. However, the description of the public body's remit in its governing law will usually be quite general and only provide a general background against which the concrete geographic information service will have to be assessed. In addition, the bylaws, policy documents and practices of the public body will give a good overview of its daily activities, but these activities are of course determined by the public body itself, without any balance check by other stakeholders[80]. So while this is a good first indication of the necessity of a particular information service, it does not provide a complete view on the opinion of society on the opportunity of this service.

Another reason for a public body to offer a particular geographic information service could be that the market fails to offer such a service, while the policymaker or the public body itself feels that there is a public need for it. There may be several reasons why the service is not provided by the private sector. For instance, the private sector may not be willing to supply the service because of its low profitability, or because market demand is insufficient. In other occasions, there may be market players willing to offer the geographic information service, but at a higher cost or lower productivity rate than the government would prefer[81].

[77] Communities and Local Government Committee (2008). *Ordnance Survey. Fifth Report of Session 2007-08*, http://www.publications.parliament.uk/pa/cm200708/cmselect/cmcomloc/268/26802.htm (accessed on 17/04/2011).

[78] K. Janssen (2010). *The Availability of Spatial and Environmental Data in the European Union. At the Crossroads between Public and Economic Interests.* Alphen a/d Rijn: Kluwer Law International.

[79] Wetenschappelijke Raad voor het Regeringsbeleid (2000). *Het borgen van publiek belang.* The Hague: Sdu Uitgevers [in Dutch].

[80] K. Janssen (2010). *The Availability of Spatial and Environmental Data in the European Union. At the Crossroads between Public and Economic Interests.* Alphen a/d Rijn: Kluwer Law International.

[81] Ibid.

However, this does not always mean that the public sector should provide the service itself. There may be other means of remedying the market failure, such as providing subsidies or regulating the private sector's behaviour. From the market failure perspective, only if it is deemed necessary in the public interest, should the public body endeavour to provide the geographic information service as part of its public task.

This public interest can be translated into several objectives, such as ensuring that the citizens know their rights and obligations, guaranteeing the neutral character of a particular geographic information service or the stimulation of diversity[82]. In general, one could state that the public interest with regard to information held by the public sector lies in obtaining accessible information from government on topics that influence the life of the citizen, e.g. legal information, but also information about the environment, traffic, public spending, or the weather. Accessibility holds different aspects: the information has to be physically accessible; it has to be financially affordable; and it should be intellectually accessible[83]. This final characteristic will entail in many cases that a public body cannot suffice with making geographic data available to the public, but that it has to add value to the data to provide an understandable and useful geographic information service. At this point, the private sector may start to feel threatened by the added-value services that are being offered by the public sector, while the public sector believes it is only fulfilling its task towards the public by providing the information service.

The previous paragraphs referred to geographic information services offered by public bodies because they believe that it is part of their public service to do so. However, these public bodies may also take their activities a step further and offer information services not because they feel they should do so in the public interest, but because they see a business opportunity. They may see these services as a chance to supplement their budget with some extra funds, or as a promotional activity that gives them more visibility with the citizens. While competition rules protect the private sector from possible market distorting behaviour from the public bodies, the question remains whether a public body should dedicate any effort and means to the creation of such 'unnecessary' geographic information services.

So how should the policymaker and the public bodies deal with this dilemma? First, they should determine whether there is a public need for the geographic information service they intend to provide. Hence, they need to consider whether there is a public interest that needs to be fulfilled by the creation of the service. Two elements need to be examined: if there is a public interest, and if so, if the public interest is not already being fulfilled by a private sector information service that is of sufficient quality and availability to the citizens to guarantee the public interest.

If there is no public interest, then government should refrain from providing a purely commercial service, but rather focus on its public remit. This would avoid any possible competition debates, and take away any possible discussions on market-distorting cross-subsidy situations. If there is a public interest, market failure may be an important indicator: if the market is already providing the geographic information service, the public body should be more reticent and question first whether an additional service is needed next to

[82] I. Baten and G. Van der Starre (1996). *Elektronische toegankelijkheid van overheidsinformatie*. The Hague: Rathenau Instituut [in Dutch]; B. Nouel (1996). "Commercialisation des données publiques et concurrence". In European Commission (ed.). *Proceedings of the European Commission Conference: Access to Public Information. A Key to Commercial Growth and Electronic Democracy*. Stockholm: European Commission, 1-9 [in French].

[83] A. Beers (1996). "Openbaarheid van overheidsinformatie". In Baten, I. and Van der Starre, G. (ed.). E*lektronische toegankelijkheid van overheidsinformatie*. The Hague: Rathenau Instituut, 51-77 [in Dutch].

the commercial services already available. If there is no such service, government should consider whether other measures, such as subsidies or the introduction of a universal service requirement[84], could be sufficient to fill the information gap. If a new service is still considered necessary, it should be carefully examined what its impact on the market will be and whether this impact is justified by the public interest that is being protected.

If the public body can determine by itself whether or not it should provide a particular service in the public interest, there is a risk that it overestimates the need for the geographic information service or underestimates the market impact the new service might have. Therefore, it is essential that stakeholders from the market, civil society and interest groups can voice their opinion on the service in question. Either this could be done in an *ex ante* evaluation process, where the public body's plans for a new geographic information service are presented to the stakeholders, or in *ex post* procedure, where the stakeholders can challenge the service if they feel it does not represent a public interest or that its negative impact on the market is too strong. While the former has the advantage that a balanced decision is taken before the service is launched, it requires a heavy administrative procedure that may slow down the response to the public need for the information service. On the other hand, in the case of a possible challenge after the service has been launched, it may be difficult to turn back the decisions made by the public body and to undo the effect on society or the market.

Conclusion

In this chapter, we considered the role of the public sector on the geographic information market. Two main issues were discussed. First, we addressed the role of the public sector from the point of view of competition law, establishing that when a public body provides a particular geographic information service as an economic activity, there are several ways in which it can distort the information market and infringe European competition law. Second, we questioned whether the public sector should even consider offering information services based on its geographic data, be it inside or outside their public task or remit. While the first issue may already cause difficulties in interpreting whether a particular geographic information service is an economic activity (and if it constitutes e.g. an abuse of dominance or illegal State aid), the second question is even more difficult.

Determining the opportunity of the provision of a geographic information service by a public body requires insight in the needs of the public, an assessment whether the public interest will benefit from the information service, and an evaluation of the possible impact of the information service on the market. Many of the public geographic data providers do not have much experience in making these assessments and making the reasons behind their decisions on their information services explicit. They will need guidance from central government on how to do this, and particularly on how to involve the stakeholders in the

[84] The concept of universal services refers to services that are considered so essential by society that everyone should have permanent access to them against reasonable conditions, both from the perspective of quality and from a financial point of view. The government may compel one or more suppliers to offer these services, because it deems this beneficial for society as a whole (see P. Valcke (2004). *Digitale diversiteit. Convergentie van media-, telecommunicatie- en Mededingingsrecht*. Brussel: Larcier [in Dutch]).

process. In its turn, central government can draw inspiration from other sectors where comparable issues have arisen, such as the sector of legal information services or public broadcasting[85], or even the telecommunications or utilities sector.

[85] Possible inspiration can be found in the description of the public value test by Karen Donders and Tim Raats in the next chapter.

Measuring Public Value with the Public Value Test: Best or Worst Practice?

Karen Donders[1] *and Tim Raats*[2]

Introduction

From spatial data to public broadcasting

The public sector has traditionally been the main producer of geographic data. It uses these data for preparing, implementing and evaluating its policy and for supporting service delivery to the citizens. However, there have also been growing trends in some Member States of the European Union towards the provision of commercial services by public sector data providers directly towards the end user, in competition with private competitors on the market. This has led to questions on the role of the public and the private sector on the information market, and to debates on how far the provision of information by the public sector should go and what should be left to the private sector[3].

This discussion was brought to the forefront by the adoption of the 2003 European Directive on the re-use of public sector information[4], and the 2007 Directive on an infrastructure for spatial information in Europe (INSPIRE)[5], which are expressly based on the concept of the public task. Until now, no agreement can be found on the roles of the public and the private sector in the spatial information market.

However, this market is not the only one that has had to deal with this debate. Other sectors have also been struggling with the division of roles between the public and the private sector. One of these domains is public broadcasting. In this chapter, we will examine how the relationship between the public broadcaster and its private counterparts is being tackled, by using the example of the United Kingdom's Public Value Test. This test can act as an inspiration for the public task debate in the market for geographic information services.

The question of the public service remit

In the 1920s and 1930s Western European countries decided to assign broadcasting a monopoly status. Although this monopoly status ended in the 1980s, public broadcasters such as the BBC, the German ARD and France Télévision still occupy important market

---

1 Center for Studies on Media Information and Telecommunication, and Institute for European Studies, Vrije Universiteit Brussel, kdonders@vub.ac.be.

2 Center for Studies on Media Information and Telecommunication, Vrije Universiteit Brussel, traats@vub.ac.be.

3 K. Janssen (2010). *The Availability of Spatial and Environmental Data in the European Union. At the Crossroads between Public and Economic Interests.* Alphen a/d Rijn: Kluwer Law International.

4 European Parliament and Council (2003). Directive 2003/98/EC of 17 November 2003 on the re-use of public sector information, *OJ L* 245, 90.

5 European Parliament and Council (2007). Directive 2007/2/EC of 14 March 2007 establishing an Infrastructure for Spatial Information in the European Community (INSPIRE), *OJ L* 108, 1.

positions. They aim to deliver public service goals and move into new media markets delivering services which go beyond traditional radio and television programmes.

The shift from so-called public service broadcasting to a public service media policy project provokes diverging reactions. Some claim the opportunities of the new media environment are not adequately grasped by the public broadcasters[6]. They should expand their activities on the Internet and mobile platforms in order to enhance the development of a public sphere in which all citizens can participate, regardless of their 'profitability' for media companies. Others – notably private media companies – reject such a view alleging public broadcasters' service delivery should be heavily constrained in the digital age, in order to protect private investments, creativity and, ultimately, innovation[7].

Until the 1980s, national (and in some cases, regional) governments autonomously decided on the organisation of public service broadcasting. They defined the task (or so-called 'public service remit') of public broadcasters, decided on funding modalities and monitored the performance of public broadcasters. Given the liberalisation of the broadcasting market in the late 1980s (by means of the 1989 Television without Frontiers Directive[8]), this situation has come to an end. Although Member States are still exclusively competent in the area of public service broadcasting, they have to accept that competition rules – which are being enforced in all sectors of economic activity – have a say in the future course of public service broadcasting policy[9].

The European Commission, which enforces the competition rules in first instance, has been confronted several times with private sector concerns about the expanding role of public broadcasters. Not surprisingly, private media companies regard this expansion as unfair competition and have filed numerous complaints with the Directorate General for Competition, alleging that the new media activities of public broadcasters should be curbed. Needless to say, public broadcasters and Member States are far from happy with the intervention of the European Commission, which is said to subordinate cultural to economic policy objectives[10].

In order to reach some sort of compromise between the apparent diverging positions on this issue, the European Commission has asked several Member States to design and implement a test for public broadcasters' new media services[11]. This test has been modelled

---

[6] G. Murdock (2004). "Building the digital commons: public broadcasting in the age of the Internet". Paper presented at the 2004 Spry Memorial Lecture, Montreal, 22 November; G. Ferrell Lowe (2010). "Beyond altruism: why public participation in public service media matters". In Ferrell Lowe, G. (ed.). *The public in public service media*. Göteborg: Nordicom, 9-36.

[7] ACT, AER, ENPA, EPC and VPRT (2009*). Joint letter to Commissioner Kroes on the revision of the Broadcasting Communication*, unpublished.

[8] Council (1989). Directive 89/552/EEC of 3 October 1989 on the coordination of certain provisions laid down by Law, Regulation or Administrative Action in Member States concerning the pursuit of television broadcasting activities, *OJ L* 298, 23.

[9] K. Donders (2010). "Europese Staatssteunregels en de openbare omroep: acties van de Europese Commissie en de Europese gerechtshoven tussen 1992 en 2009". *Mediaforum* 9, 270-278 [in Dutch]; M. Wheeler (2010). "The European Union's Competition Directorate: State aids and public service broadcasting". In Iosifidis, P. (ed.). *Reinventing public service communication: European broadcasters and beyond*. Basingstoke: Pallgrave Macmillan, 49-62; K. Donders and C. Pauwels (2008). "Does EU policy challenge the digital future of public service broadcasting? An analysis of the Commission's State aid approach to digitization and the public service remit of public broadcasting organizations". *Convergence. The International Journal of Research into New Media Technologies* 14(3), 295-311.

[10] K. Donders and C. Pauwels (2008). "Does EU policy challenge the digital future of public service broadcasting? An analysis of the Commission's State aid approach to digitization and the public service remit of public broadcasting organizations". *Convergence. The International Journal of Research into New Media Technologies* 14(3), 295-311.

[11] K. Donders and C. Pauwels (2010). "The introduction of an ex ante evaluation for new media services: 'Europe' asks it or public broadcasters need it?". *International Journal of Media and Cultural Politics* 6(2), 133-148.

after the British 'Public Value Test'. It is a tool that evaluates *ex ante* whether a proposed new media service lives up to pre-set public service objectives and does not overly distort the market. It is a balancing act between more elevated public interest objectives on the one hand and down to earth market logics on the other hand.

At present, Member States (including Germany, Belgium, the Netherlands and Ireland) are complying with the European Commission's request. Opponents claim the test is a bureaucratic monster that is overly expensive and can hardly be expected to objectify essentially political decisions on the scope of government intervention in the market[12]. Proponents argue a test is 'natural', as all services of public broadcasters should exhibit sufficient public value and, in doing so, make distortions of the market admissible[13].

This chapter studies the British Public Value Test, in order to evaluate whether it is indeed a tool that can find a balance between public and market interests. After this introduction, the chapter consists of six parts. In part two the evolution from public service broadcasting to public service media and, in addition, the increasing need to demarcate public from commercial service delivery is addressed. Part three briefly discusses European State aid policy as a basis for asking Member States to implement a public value test. After that, the main aspects of the British Public Value Test and its implementation in practice are addressed. Sixthly, the results of the British Public Value Test are evaluated. Finally, a number of conclusions and policy recommendations – also for other sectors in which competition issues are at play – are outlined.

From public service broadcasting to public service media: how defining the public service remit became a key issue

Since its inception in the 1920s, the need to define more clearly what public broadcasting is and demarcate public broadcasters' public from commercial services has continued to increase. In what has often been described as the analogue era, media regulation was solely based on traditional broadcasting means, *i.e.* radio and television. Spectrum scarcity served as a decisive argument for installing a broadcasting monopoly in Western Europe, with public service broadcasters being funded referring to the rather vague Reithian[14] adage 'inform, educate, entertain'. In the 1980s the liberalisation of broadcasting markets invoked fundamental changes on all levels of public broadcasting organisations. The latter were now *de facto* competing with commercial broadcasters. Nevertheless, media markets were still restricted mainly to traditional broadcasting companies. In recent years however, the convergence of media technologies gave rise to a multitude of new actors, shifts in market structures, changing consumer behaviour and adaptations of media regulation. All of this has brought about a series of transitions affecting the outlook, financing and structure of public broadcasters. More importantly, the current environment requires a more evidence based definition of public broadcasters' necessity. Should we define public broadcasters' public service remit on the basis of an observation of market failure or do we set out from

---

[12] J. Bardoel and M. Vochteloo (2009). "Media policy between Europe and the nation-state: the case of the EU Broadcast Communication 2009". Paper presented at the ECREA-CLP Workshop on 'new direction for communication policy research', Zürich, 6-7 November.

[13] T. Suter (2008). "Setting public goals, defining public value, securing public support – the challenge for Europe's PSBs". Presentation at the conference on 'Public service media in the digital age', Strasbourg, 17 July.

[14] John Reith was the first Director General of the BBC.

democratic and social values?[15] Indeed, abundance of choice and the extension of public broadcasters' activities into the online – digital – realm, have called for a clearer description of public broadcasters' public service remit and, in general, a more clearly spelled-out justification of public broadcasters' activities.

Ideological imperatives

Defining the public remit and providing justification of public broadcasters' activities has proven to be more difficult than perhaps initially expected. Public broadcasters' public service remit in continental Europe was and still is more often than not based on strong ideological motives. Traditionally, public broadcasting organisations were characterised as highly politicised institutions serving mostly political rather than public causes. Even today, politicisation remains a key concern when dealing with public broadcasting policy. Furthermore, the public broadcasters remain embedded in regulatory systems that are not adapted to the current media environment, which transcends television and radio broadcasting. In light of the outdated regulation of public broadcasters in the new media environment, the need for re-defining and demarcating the public service broadcasters' remit has been mostly resounded from an economic angle, and is mostly being called upon by public broadcasters' competitors[16].

From a more social-democratic angle, the concept of public service media can be seen as an attempt to, firstly, re-position broadcasters in a digital ecology and better meet their existing remit by acclaiming the online domain and the delivery of digital services as crucial[17]. Secondly, next to shaping public broadcasters' activities, it can also be seen as a legitimising instrument for public broadcasters' activities in a digital context, with public broadcasters transposing the very arguments that provided legitimacy in the analogue era to the digital era[18]. In substantially defining and legitimising their activities – setting out from the public service media concept – they aspire to counter the criticism and external pressures from various stakeholders.

Social-democratic perspectives start from a set of common assumptions. Firstly, public service broadcasting and, in extension, public service media attribute to the democratic

[15] M. Tracey (1998). *The decline and fall of public service broadcasting*. Oxford: Clarendon Press; J. Bardoel and L. d'Haenens (2008). "Public service broadcasting in converging media modalities. Practices and reflections from the Netherlands". *Convergence. The International Journal of Research into New Media Technologies* 14(3), 351-360; K. Jakubowicz (2007). *Public Service Broadcasting: a New Beginning, or the Beginning of the End,* http://www.coe.int/t/dghl/standardsetting/media/Doc/PSB_Anewbeginning_KJ_en.pdf (accessed on 10/02/2008); R. Collins et al. (2001). "Public service broadcasting beyond 2000: Is there a Future for Public Service Broadcasting?." *Canadian Journal of Communication* 26(1), http://www.cjc-online.ca/viewarticle.php?id=615 (accessed on 10/08/2011); P. Lee-Wright (2008). "Virtual News: BBC News at a 'future media and technology crossroads'". *Convergence. The International Journal of Research into New Media Technologies* 14(3), 249-260.

[16] J. Bardoel and G. Ferrell Lowe (2007). "From public service broadcasting to public service media. The core challenge". In Ferrell Lowe, G. and J. Bardoel (ed.). *From public service broadcasting to public service media.* Göteborg: Nordicom, 9-28.

[17] R. Collins et al. (2001). "Public service broadcasting beyond 2000: Is there a Future for Public Service Broadcasting?". *Canadian Journal of Communication* 26(1), http://www.cjc-online.ca/viewarticle.php?id=615 (accessed on 10/08/2011).

[18] G. Murdock (2005). "Building the digital commons. Public broadcasting in the Age of the Internet". In Lowe, G.F. and Jauert, P. (ed.). *Cultural Dilemmas in Public Service Broadcasting*. Göteborg: Nordicom, 213-229; H. Moe (2008). "Dissemination and dialogue in the public sphere: a case for Public Service Media Online". *Media, Culture & Society* 30(3), 319-336.

needs of society. Secondly, the 'old' public broadcasting organisations are still the best and also most well-known instrument to provide public media services. Thirdly, economic market failure perspectives depart from a highly utopian conception of society and the market and should, for that reason and according to proponents of the social-democratic perspective, not be taken into consideration in the development of public broadcasting policy. Fourthly, technological evolutions do not make a public broadcasting system superfluous. However, they necessitate a fundamental rethinking of public broadcasters and their remit. Fifthly, and continuing along the line of technology neutrality, public broadcasters' public service remit is not exclusively tied to specific distribution platforms. Finally, public broadcasters' public service remit should be defined in a broad way and needs to comprise of a multitude of services, genres and platforms[19].

The public service media concept

At the core of what public broadcasters should do, according to public service media perspectives, is the holistic remit of public broadcasters. Looms describes public service media as 'all things – all people – on all platforms, anytime'[20]. While a simplification of public service media would do the concept harm, Looms' short description markedly portrays its key characteristics. The public service media concept basically entails a public broadcaster that performs on all platforms and distribution media in the provision of its content and public services. The idea of public service media presupposes a holistic broadcaster that has to remain popular if it wants to be effective and that provides a mix of genres and programming that appeal to mass audiences as well as diverse target groups. It sets out from existing public service broadcasting values and strengths such as quality, innovativeness and creativity, moral accountability and trust, universality and distinctiveness, diversity, impartiality and loyalty[21].

The main argument put forward in favour of public service media is the need for multimedia and cross-media provision of content and services in order to fully reach the

---

[19] H. Moe (2008). "Dissemination and dialogue in the public sphere: a case for Public Service Media Online". *Media, Culture & Society* 30(3), 319-336; J. Bardoel and G. Ferrell Lowe (2007). "From public service broadcasting to public service media. The core challenge". In Ferrell Lowe, G. and J. Bardoel (ed.). *From public service broadcasting to public service media.* Göteborg: Nordicom, 9-28; M. Aslama and T. Syvertsen (2007). "Public service broadcasting and new technologies: marginalisation or re-monopolisation". In De Bens, E. (ed.). *Media between culture and commerce.* Bristo: Intellect, 167-179; A. Leurdijk (2007). "Public service media dilemmas and regulation in a converging landscape". In Ferrell Lowe, G. and Bardoel, J. (ed.). *From public service broadcasting to public service media.* Göteborg: Nordicom, 71-86; J. Steemers (2003). "Public Service Broadcasting Is Not Dead Yet: Strategies in the 21st Century". In Lowe, G. and Hujanen, T. (ed.). *Broadcasting and Convergence: New Articulations of the Public Service remit.* Göteborg: Nordicom, 123-136; S. Barnett (2002). "Which end of the telescope? From market failure to cultural value". In Cowling, J. and Tambini, D. (ed.). *From public service broadcasting to public service communications.* London: IPPR, 34-45; C. Leys (2001). *Market-driven politics: neoliberal democracy and the public interest.* London: Verso; M. Tracey (1998). *The decline and fall of public service broadcasting.* Oxford: Clarendon Press; A. Brown (1996). "Economics, public service broadcasting, and social values". *The Journal of Media Economics* 9(1), 3-15; N. Garnham (1990). *Capitalism and communication: global culture and the economics of information.* London: Sage Publications.

[20] P.O. Looms (2006). "Public service media: all things to all people – on all platforms, anytime?". In Nissen, C.S. (ed.). *Making a difference: public service broadcasting in the European landscape.* Eistleigh: John Libbey Publishing, 95-114.

[21] S. Barnett (2006). "Public service broadcasting: a manifesto for survival in the multimedia age (a case study of the BBC's new charter)". Paper presented at the RIPE Conference, Amsterdam, 2-6 October.

audiences; in other words, public broadcasters need to go where their audiences take them[22].

Jakubowicz and others even argue that the transition into the online realm and the embracement of public service media gives public broadcasters the opportunity for the very first time to truly cover the whole of their public service remit[23]. Moreover, the argument for public broadcasting organisations to be active online is enforced because of commercial pressures, since large profit-chasing companies have also appeared in the online environment. Public service broadcasting in that sense, offers a public counterweight for commercial interests:

> "[O]nline media produced by public service broadcasters play an important role in counterbalancing dominant trends such as concentration and commercialization. Attempts to limit public service broadcasting activities to radio and broadcasting would eliminate these online voices to the detriment of competition in quality"[24].

More concretely, public broadcasters' provision strategy or 'full-portfolio strategy' should include the web and digital services, on-demand content, thematic digital channels, mobile, etc. Although often contested by private players, within a public service media scenario the following services can be offered by public broadcasters in the digital environment: alert-services on mobile phones, informative services through text messages, web- and sms-voting, forums and communities on websites, on-demand archive services, personal video guides, operated by the public broadcaster, streaming of radio and television programmes, online games, interactive services for entertainment shows, etc. All of the above services should be offered besides traditional public service radio and television programmes[25].

This plethora of services requires a full-fledged adaptation of public broadcasters' current activities. Public broadcasters are challenged to provide all these services at the same time, without neglecting their basic public service mission[26]. This, again, brings us to the basic question of how to define the public service remit and how to demarcate public from commercial services. Scholars provide lists of services they think fit the public service remit. However, little legal certainty flows from the more ideological statements made within the social-democratic perspective.

While today, no regulatory authority – including the European Commission – questions the presence of public broadcasters in the online domain[27], the scope and contents of online public service activities are under constant scrutiny and are being increasingly criticised by public broadcasters' opponents as well as regulatory bodies. Although this appears contradictory to the apparent widespread acceptance of the public service media

---

[22] I. Banerjee and K. Seneviratne (ed.) (2005). *Public Service Broadcasting: A Best Practices Source Book.* Paris: UNESCO.

[23] K. Jakubowicz (2007). *Public Service Broadcasting: a New Beginning, or the Beginning of the End,* http://www.coe.int/t/dghl/standardsetting/media/Doc/PSB_Anewbeginning_KJ_en.pdf (accessed on 10/02/2008).

[24] J. Trappel (2008). "Online media within the public service realm? Reasons to include online into the public service mission". *Convergence. The International Journal of Research into New Media Technologies* 14(3), 313-322.

[25] K. Jakubowicz (2007). *Public Service Broadcasting: a New Beginning, or the Beginning of the End,* http://www.coe.int/t/dghl/standardsetting/media/Doc/PSB_Anewbeginning_KJ_en.pdf (accessed on 10/02/2008).

[26] OFCOM (2004). *OFCOM review of public service broadcasting: Phase 1 – Is television special?,* http://stakeholders.ofcom.org.uk/consultations/psb/ (accessed on 10/08/2011).

[27] J. Bardoel and L. d'Haenens (2008). "Public service broadcasting in converging media modalities. Practices and reflections from the Netherlands". *Convergence. The International Journal of Research into New Media Technologies* 14(3), 351-360.

concept with supporters of public broadcasters, criticism essentially flows from the gaps in the public service media concept and, moreover, from the competitive situation in which public broadcasting organisations find themselves. Economic downfall and collapsing advertising markets have confronted public service media organisations with large-scale budgetary reforms and cutbacks, but they have posed even more severe restrictions on public broadcasters' competitors. In such a climate, complaints against the 'stable' funding of public broadcasters can be easily explained.

European State aid policy and the 'Amsterdam test'

The funding of public broadcasters has often been questioned under the European Union's State aid policy. Government mechanisms that grant a financial advantage to a particular sector or a specific company (like a subsidy or a tax exemption) are considered State aid and are in principle prohibited in the European Union. In order to prevent subsidy wars between Member States, which would harm the internal market, State aid is subject to strict rules enforced in first instance by the European Commission. State aid rules have been applied to the funding of public service broadcasting since the beginning of the 1990s, when private broadcasters filed complaints with DG Competition, alleging public broadcasters behaved unfairly in broadcasting markets. Since then, the European Commission has attempted to develop a consistent approach with regard to the issue of public broadcasters expanding activities to new media markets.

Confronted with allegations of unfair competition in relation to the launch of thematic channels like Phoenix (a German documentary channel) or BBC News 24 (a British information channel), the European Commission confirmed Member States' right to define the public broadcasting remit at the end of the 1990s. It was up to the Member States to decide whether certain channels fit within the public service remit or not. This approach was based on the 1998 *Amsterdam Protocol*. This Protocol, which is annexed to the EC Treaty, recognises the invaluable contribution that public broadcasting organisations make to cultural, social and democratic objectives of society. The Protocol explicitates that the Member States are fully competent in determining the funding and regulation of their public broadcasters. However, any funding should not adversely affect trade and competition in the Community. In that sense, the Protocol grants a special status to public service broadcasting, but does not exempt it from the application of the State aid rules[28].

At first: co-defining the public service remit

According to these rules, funding schemes are permissible in case they comply with a number of good governance principles. For public broadcasting, three principles are relevant. Firstly, the public service remit must be defined in a clear way. There should be no doubt about the public or commercial nature of services. Secondly, the task of public broadcasters must be officially entrusted to the organisations delivering public broadcasting services. This entrustment is to be accompanied by objective and independent control of the fulfilment

[28] I. Nitsche (2001). *Broadcasting in the European Union: the Role of Public Interest in Competition Analysis*. The Hague: TMC Asser Press.

of the public service obligations. Thirdly, aid must be proportional. In other words: the net cost of a public broadcaster should be determined, and government support cannot exceed this cost[29]. When the European Commission investigates the funding of public broadcasters, these three principles guide the investigations and assessment.

On the basis of these principles (listed in the 2001 Broadcasting Communication[30]), a number of funding schemes were investigated and evaluated. In all investigations related to new media, controversy arose about the legitimate scope of Commission interference with the public service remit. Whereas the Amsterdam Protocol emphasises Member States' right to define the remit, case practice after 2001 indicated that the European Commission attempted to co-define the public service remit on the basis of the three principles that guide its State aid investigations into the funding of public service broadcasting. The Commission demanded, for example, that any new service developed by a public broadcaster remained closely associated with its traditional television and radio programmes. Public subsidies could only be used to support services that were linked to public broadcasters' core activities[31]. This approach obviously went against public service media logic as it made a distinction on the basis of technology (as Internet or mobile services were deemed outside the core of public broadcasters' public service delivery) and service format (traditional programs being supposedly a more legitimate part of public broadcasters' activities). Moreover, it infringed Member States' autonomy to define the remit[32]. In addition, DG Competition stressed that new services should possess some added public value in comparison with already existing offers of public broadcasters[33]. Again, the approach failed to see new services as a fully legitimate part of public broadcasters' services portfolio.

Subsequently: urging Member States to define the remit

Confronted with opposition against DG Competition co-defining the public service remit, the competition authorities changed tactics and started urging Member States to effectively regulate public broadcasters' activities from 2007 onwards. In other words, DG Competition was no longer prepared to accept that public broadcasters themselves determined which (new) activities were covered by their remit. Unrestricted expansion strategies in the new media environment caused by a hands-off approach of national governments were thus no longer permissible. A clearly defined remit had to be developed by Member States – and thus not by the Commission or public broadcasters – in order to ensure transparency and predictability for other players in the market[34].

---

[29] A. Antoniadis (2006). "The financing of public service broadcasting". In Rydelski, M.S. (ed.). *The EC State aid regime: distortive effects of State aid on competition and trade*. London: Cameron May, 591-630; A. Bavasso (2006). "Chapter 17: Broadcasting". In Hancher, L., Ottervanger, T. and Slot, P.J. (ed.). *EC State Aids*. London: Thomson Sweet and Maxwell, 420-438.

[30] European Commission (2001). Communication on the application of State aid rules to public service broadcasting, *OJ C* 320, 5-11.

[31] European Commission (2003). BBC Digital Curriculum Decision. N37/2003, C(2003)3371fin.

[32] V. Wiedemann (2004). *Public service broadcasting, State aid and the Internet: emerging EU law*, http://www.ebu.ch/en/union/diffusion_on_line/pub_broadcasting/ (accessed on 05/08/2007).

[33] European Commission (2007). Decision: Licence fee funding of German public broadcasters ARD and ZDF. E3/2005. *OJ C* 185, 1.

[34] European Commission (2008). Decision: annual funding of Flemish public broadcaster VRT, E8/2006. *OJ C* 143, 7.

Instead of co-defining the remit (which is against the division of competences in the field of State aid and public services), the European Commission limited itself to one question: is there a mechanism, imposed by Member States, that ensures that public broadcasters are active in these markets where they are required to be, or do Member States uphold a hands-off approach vis à vis public broadcasters? On the basis of this question, DG Competition for instance asked the Flemish, German and Irish governments to develop a mechanism that ensures that new services are covered by the remit. In the Commission's opinion, an *ex ante* test is a suitable mechanism to cope with legally ambiguous situations in which dating websites, video games, chats and other Internet services are offered on the basis of public funding. DG Competition has introduced this mechanism (dubbed 'Amsterdam test') in case practice and, moreover, in its recent update of the rules applying to the funding of public service broadcasting[35].

In order to ensure that new media services effectively fit within the public service remit, the Commission thus encourages Member States to introduce a two-sided ex ante test. On the one hand, new media services' public value must be evaluated. On the other hand, their positive and negative effects on the market should be calculated as well[36]. On the basis of these two 'legs', governments should approve the delivery of a new service or not. Public broadcasters thus require prior approval for 'new' media services. The Member States have to determine themselves what a new media service is. Examples of new services are Internet streaming or on-demand services like the BBC iPlayer or even news websites like heute.de.

In short, the European Commission wants to prevent that a number of commercial services are swept away under the carpet of democracy, social cohesion, universal service delivery, cultural diversity, etc[37]. For that purpose, the two-sided test described above is suggested as an adequate means to ensure that public broadcasters can do their job while not preventing others from doing theirs.

Main aspects of the British Public Value Test

What does an ex ante test look like in practice? Is it really a best practice of enhancing media governance and was the European Commission, hence, right in enforcing it in Member States' public broadcasting policies? We start our analysis with a brief discussion of the main aspects of the British Public Value Test.

---

[35] The Amsterdam Protocol assigns Member States the right to define and organise their system of public broadcasting provided that the system does not infringe competition rules to an unacceptable extent. The ex ante test proposed by the European Commission copies this approach. The test evaluates whether a service is effectively pursuing public interest objectives and is in doing so not adversely or unnecessary distorting the market. Given Member States' and public broadcasters' support for the Amsterdam Protocol, the European Commission dubbed the ex ante test 'Amsterdam test'. This semantic attempt to raise support for the European Commission initiatives did not lower Member States' resistance against the Commission's proposals to revise the State aid rules for public service broadcasting, however. The rules concerning the funding of public broadcasting organisations were updated in the beginning of July 2009.

[36] European Commission (2009). Communication on the application of the State aid rules to public service broadcasting, *OJ C* 257, 1.

[37] K. Donders and C. Pauwels (2009). "European State aid rules and the public service broadcasting remit in the digital age: analyzing a contentious part of European policy and integration". In Garcia-Blanco, I., Van Bauwel, S. and Cammaerts, B. (ed.). *Media agoras: democracy, diversity and communication*. Cambridge: Cambridge Scholars Publishing, 178-197.

The Public Value Test has been used from 2007 onwards for a total of four BBC services: (1) the BBC iPlayer, (2) the launch of a high definition channel, (3) a Gaelic television channel, and (4) local news video services. These services are so-called significant changes in the public service delivery of the BBC. For such services, a prior approval of the BBC Trust (a control body internal to, but independent from the BBC) is required[38]. The test can be considered as a dynamic addition to the (in contrast static) Royal Charter. This contract between the UK's government and the BBC provides the main rules and objectives for the public broadcasting organisation. Its renewal, every ten years, is usually preceded by an extensive revision and consultation process. Because of the static character of the Charter (which is beneficial from a legal certainty point of view, both for the BBC and its competitors), a dynamic instrument to assess whether certain, not yet envisaged, new activities fit within the public service remit of the BBC was found desirable and also necessary in the fast-changing media environment. The instrument must enable the BBC to modify its offer, while at the same time ensuring that significant modifications remain in fact covered by the Royal Charter. The Public Value Test is thus not regarded as a tool that limits the public broadcaster's activities, but rather as an instrument that provides solid ground for the expansion of activities within the public service remit of the BBC.

The Public Value Test consists of two parts: a Public Value Assessment and a Market Impact Assessment, conducted by the BBC Trust and Ofcom respectively. The BBC Trust coordinates the entire process of balancing public value (the part for which it is responsible itself) and the possible positive and negative market impact of a service[39]. The Public Value Test is a tool to balance public broadcasters' obligation to fulfil the public service remit, also in new media environments, with private competitors' concerns about such an expansion. At the same time, it aims to ensure that the BBC's licence fee supports public and not commercial services. All new services should serve the needs of individual licence fee payers, society as a whole (meaning: the service must serve the public purposes[40] of the BBC), and represent value for money. A value for money logic, one of the core principles of new public management theory, is as such introduced in the functioning of public broadcasters. Although the Department for Culture, Media and Sport supports the conduct of a Public Value Test, it is not directly involved in the entire process. This was different before 2007, when the Department was involved in similar exercises.

The BBC itself has repeatedly stressed the importance of measuring public value. Private competitors' interests were also considered, but with its document *'Building public value',*

---

[38] BBC Trust (2007). *Public Value Test (PVT): Guidance on the conduct of the PVT*, http://www.bbc.co.uk/bbctrust/assets/files/pdf/regulatory_framework/pvt/pvt_guidance.pdf (accessed on 10/08/2011); M. Michalis (2010). "The British Public Value Test: Structure and implementation". Presentation at the IES & IBBT-SMIT Conference Exporting The Public Value Test: Assessing EU Member States' evaluation of public broadcasters' new media services, Brussels, 25 May; D. Gleeson (2007). "The governance of the new BBC". Presentation to the BBC International Day. London, 7 September.

[39] BBC Trust (2007). *Public Value Test (PVT): Guidance on the conduct of the PVT*, http://www.bbc.co.uk/bbctrust/assets/files/pdf/regulatory_framework/pvt/pvt_guidance.pdf (accessed on 10/08/2011).

[40] These public purposes are listed in the Royal Charter. There are six public purposes: (1) Sustaining citizenship and civil society; (2) Promoting education and learning; (3) Stimulating creativity and cultural excellence; (4) Representing the UK, its nations, regions and communities; (5) Bringing the UK to the world and the world to the UK; and (6) Delivering to the public the benefit of emerging communications technologies and services. In every public value assessment, the Trust scrutinises whether a service meets all these purposes or the purposes the BBC management's proposal focuses on.

the BBC mainly focused on the public service aspect of new media service delivery[41]. The Public Value Test was considered an appropriate tool – because of far less government intervention – to measure public value in the process of the BBC building it. In addition, other BBC services are also regularly assessed by the BBC Trust *vis à vis* their public value[42]. So, the regime of assessment or the pervasiveness of auditing within the BBC is not limited to new media services only[43].

Implementation of the Public Value Test

Since its inception in 2007, the Public Value Test has been applied to four services. The design of the BBC Trust's Public Value Assessments was similar in all four tests. The BBC Trust evaluated to what extent the BBC's proposal would fit with the BBC's public purposes (*cf. supra*). Subsequently, it assessed the quality and distinctiveness, the impact, reach and value for money of the new (or significantly altered) service. Ofcom's Market Impact Assessment tackled a number of questions that relate to possible negative and positive effects of the BBC Management's proposals on other players in the market. A Joint Steering Group (JSG) set the terms of reference for the Market Impact Assessment. The JSG consisted of members of the BBC Trust, Ofcom and external experts. This group decided on the methods used to perform the Market Impact Assessment and the relevant markets that are included. Ofcom was thus not solely responsible for the design of the Market Impact Assessment[44]. This working method allows that the Market Impact Assessment is oriented at enabling the BBC to fulfil its remit while at the same time ensuring fair competition.

Following Gibbons' comments on Ofcom's predominantly market and competition oriented policies[45], the current working method for the Market Impact Assessment aims to prevent that competition policy takes over from sector specific public broadcasting policies and that, as a consequence, these two different types of policy can be complementary. The assessment itself, although different for the four services tested, consisted of two steps. Firstly, the relevant markets were defined: which services were likely to be affected by the BBC's proposal? Gareth Davies, Project Director of the Market Impact Assessment with Ofcom, has stressed that market definitions are inspired, but not based, on competition law principles[46]. In general, so he said in relation to the BBC proposal on on-demand services, broader definitions are upheld. These take into account the specificity of public service broadcasting. Secondly, the possible impact was assessed. This impact assessment is based on forecasting methods for both short and long term reactions to the BBC's proposals, responses to Ofcom's questionnaires, etc.

---

[41] BBC (2004). *Building public value: renewing the BBC for a digital world*, http://downloads.bbc.co.uk/aboutthebbc/policies/pdf/bpv.pdf (accessed on 10/08/2011).

[42] B. Franklin (2005). "Introduction". In Franklin, B. (ed.). *Television policy: the MacTaggart Lectures*. Edinburgh: Edinburgh University Press, 1-25.

[43] R. Collins (2010). "Exporting the Public Value Test: assessing EU Member States' evaluation of public broadcasters' new media services. The United Kingdom". Presentation at the IES & IBBT-SMIT Conference Exporting The Public Value Test: Assessing EU Member States' evaluation of public broadcasters' new media services, Brussels, 25 May.

[44] C. Woolard (2008). "Background and framework". Presentation at OFCOM's Information Seminar 'Market Impact Assessment of the BBC's new on-demand proposals', London, 18 September.

[45] T. Gibbons (2005). "Competition policy and regulatory style – issues for OFCOM". *Info* 7(5), 42-51.

[46] G. Davies (2008). "Methodology for Market Impact Assessments". Presentation at the OFCOM Information Seminar 'Market Impact Assessment of the BBC's new on-demand proposals', London, 18 September.

The iPlayer test

The tests resulted in three approvals with conditions and one negative advice of the Trust. The first test concerned the delivery of on-demand services by the BBC (including services offered via iPlayer). This service was approved because it presented high public value, following technological evolutions and allowing licence fee payers to make use of BBC content on a more flexible basis. The services could potentially create a bigger market for on-demand content. However, a number of conditions were attached to the approval of the service. These conditions related nearly exclusively to concerns of the market players. On-demand services could be offered, but various time limits for different types of on-demand services were introduced in order to ensure that consumers would not expect all content (also of private undertakings) to be free of charge[47].

High definition television

The second test, concerning the launch of a high definition channel, resulted in a positive result as well. Again, the BBC Trust emphasised the need for platform neutrality as the high definition channel was only offered via the digital terrestrial television platform Freeview and not via cable and satellite (which would be desirable in light of the BBC's required platform neutrality). Overall, the Trust considered it necessary that the BBC contributed to innovation and the further spread of the high definition technology. A number of conditions accompanied the approval as well. These conditions related to both public value and market impact concerns. For example, the Trust stressed that the high definition channel should not merely offer BBC One prime time content, but should strive for a mix of genres and content suitable for the launch and promotion of a high definition channel[48]. The BBC Trust's position did not deviate from Ofcom's stance in that respect. Ofcom also insisted on a balanced offer of the channel, albeit with market related concerns in the back of its mind. Premium content, i.e. film and sports, should be limited to a maximum of 20% of the channel's schedule in order to prevent head-on competition with pay-television providers. Overall, Ofcom considered that the impact on viewing shares of commercial broadcasters would be limited. Competition between broadcasters was determined more, so Ofcom concluded, by the content offered (e.g. premium content) than picture quality[49].

Gaelic television

A Public Value Test evaluating the launch of a Gaelic digital television channel followed the high definition television test. Whereas the two previous tests focused on both public value and market impact, the third test was mainly concerned with the public value of a digital television channel for the Gaelic speaking community. Ofcom concluded that few relevant

---

47 BBC Trust (2007). *BBC on-demand proposals: Public Value Test final conclusions*, http://www.bbc.co.uk/bbctrust/assets/files/pdf/consult/decisions/on_demand/decision.pdf (accessed on 10/08/2011).

48 BBC Trust (2007). *Public Value Assessment: BBC HDTV proposal*, www.bbc.co.uk/bbctrust/assets/files/pdf/consult/hdtv/pva.pdf (accessed on 10/08/2011).

49 OFCOM (2007). *BBC new on-demand proposals – Market Impact Assessment: executive summary*, http://stakeholders.ofcom.org.uk/market-data-research/tv-research/bbc-mias/ondemand/bbc-ondemand/ (accessed on 10/08/2011).

market impact related issues (positive or negative) could be observed in relation to this BBC proposal. However, the BBC Trust seriously doubted what the public value of the service was. Only a small minority of the Gaelic community speaks the language. As such, the reach and, consequently, impact of the service (see the evaluation criteria in the Public Value Assessment discussed above) would be limited. Although the Trust considered that the channel was very important from a cultural perspective, the BBC's strategy to promote the channel was said not to be adequate to reach the defined target audience. In addition, there was not a lot of cooperation between the BBC and partners in the field of education. The Trust also questioned the cooperation between the BBC and the Gaelic Media Service (GMS), which was a partner in the project. The Trust required guarantees that the GMS would effectively fulfil its financial obligations and that the rather bureaucratic management of the GMS would not impede creativity and a smooth management of the channel. As a consequence of these provisional conclusions of the BBC Trust, GMS and the BBC re-evaluated the service and came back with a number of proposals to remedy the Trust's concerns. Subtitling, partnerships with the educational field, marketing strategies connected with specific target audiences, etc. were considered to overcome most criticism. The Trust, finally, agreed. Yet, it insisted that the service would be evaluated already in 2010 in order to see whether it indeed delivered public value[50].

Local video

The fourth Public Value Test was concluded in February 2009. It resulted in the first negative result, which was preceded and followed by considerable controversy. The test evaluated the so-called local video proposal of the BBC. The service's aim was to strengthen the BBC's coverage of local issues (mainly sports and news). It would do so by offering on the BBC's regional websites up to 10 videos a day dealing with regional topics (news, sports, culture, arts, heritage, weather, etc.). These videos would be produced locally. The BBC proposed this service as a consequence of an observed 'purpose gap' between its public purpose to serve the United Kingdom, its regions and communications on the one hand and target audiences (*i.e.* licence fee payers) on the other hand. Although the BBC Trust agreed that there was a purpose gap, it was not sure that this service would close it. The service would be offered via broadband Internet only, whereas the target audiences desiring regional news not always have access to broadband Internet and, more importantly, still prefer traditional (often print) media for regional news coverage. The Trust suggested, after taking into consideration the results of the Market Impact Assessment[51], that the BBC management present another service proposal to strengthen its regional news coverage that would effectively close the purpose gap. The negative outcome of the Market Impact Assessment played a significant role in the Trust coming to such a conclusion. Ofcom had concluded in its Market Impact Assessment that the roll-out of the local video proposal would result in negative impacts on the market. On the basis of its assessment, Ofcom argued that the sector covering regional news could face an immediate loss of up to 4% of its annual revenues. The assessment also indicated that the combination of uncertain prospects for

[50] BBC Trust (2007). *Gaelic digital services: Public Value Assessment*, http://www.bbc.co.uk/bbctrust/assets/files/pdf/consult/gaelic_digital_service/pva.pdf (accessed on 10/08/2011).

[51] OFCOM (2008). *Market Impact Assessment of the BBC's local video service*, http://stakeholders.ofcom.org.uk/binaries/research/tv-research/bbc-mias/mia-bbc-local-video.pdf (accessed on 10/08/2011).

the print sector in combination with a strengthening position of the BBC in this market would result in a serious decline of investments of the sector in regional news coverage. The sector for regional news is rather fragmented and was not considered powerful enough to cope with the BBC as a strong competitor. The lack of investment of the commercial sector would be reinforced by the economic downturn in 2008 and result in less pluralism and competition. In the opinion of Ofcom, this should not be the consequence of government intervention. Ofcom concluded its analysis with a negative advice, urging the BBC Trust not to accept the service without a strong public value finding in the Public Value Assessment[52]. Since the Public Value Assessment was indeed doubtful about the reach and impact of the local video service, the proposal was rejected. The BBC Trust made explicit that it would welcome new proposals and encouraged the BBC to further its plans to cooperate with the private sector in this regard. However, such public-private partnerships, which presumably result in efficiency gains, should be scrutinised as well as they can also distort competition[53].

Evaluation of the Public Value Test

Scientific and political evaluations of the Public Value Test are largely lacking for the moment. Nevertheless, it is important to see whether the instrument has reached its most important objectives (i.e. ensuring public value of new media offers and constraining possible market distortive effects). Five different observations can be made when analysing the four tests that have been conducted so far.

Clarity of concepts is key

Firstly, it is not always clear what a new or significantly altered service is. Few documents expand on the question why a service requires a Public Value Test. It seems that the BBC Trust and the BBC management themselves interpret the criteria that determine whether or not there is a new service or a significant alteration to an existing service. This is not *per se* problematic, but reveals that a list of flexible criteria to determine whether or not a service is new and requires a test is arbitrary to some extent. In addition, the qualification of a service as a new or a significant alteration of existing offers depends highly on the circumstances. The local video proposal, for example, could have been considered as an extension of the existing offers with regard to local news (to be accessed via bbc.co.uk). However, the service was classified as a new service, mainly because of its financial implications (the budget could amount to £68 million, which was a significant increase of the BBC's financial means in this area) and, of major importance, enormous private sector lobbying against the service. Particularly the print sector, hit hard by the economic crisis, reacted furiously against the development of the local video services proposal[54]. This does not necessarily imply that the

---

[52] Ibid.

[53] BBC Trust (2009). *Local Video: Public Value Test conclusions*, http://www.bbc.co.uk/bbctrust/assets/files/pdf/consult/local_video/decision.pdf (accessed on 10/08/2011).

[54] See e.g. V. Shaw (2009). "BBC Trust rejects online local video news service", http://www.independent.co.uk/news/media/online/bbc-trust-rejects-online-local-video-news-service-1629929.html (accessed on 5/10/2009); D. Sabbagh (2008). "BBC plans for local outlets under threat", http://business.timesonline.co.uk/tol/business/industry_sectors/media/article5201742.ece (accessed on 5/10/2009).

classification of the local video service as a new service was incorrect. It just indicates that no classification is completely objective in the sense that without the financial crisis and the structural issues the print sector is facing, the classification of local video might have been different or, even if the service was classified as a new service, the result of the test might have been positive.

The definition of 'new' services also affects the scope of services made subject to a test. The more narrow the definition, the less tests one can observe, and vice versa. In the United Kingdom, we basically observe *ad hoc* definitions of new services on the basis of a very vague list of criteria. In other countries, for example Germany, the scope of the test is much bigger and, in fact, all Internet services are being tested[55]. In Flanders, on the contrary, the scope is even more confined than is the case in the United Kingdom. The Flemish government has made a list of services that are not new. This list is so wide that it seems impossible to imagine which service could actually qualify as new[56].

The definition of a 'new service' has provoked a lot of discussion. Some stakeholders allege that the criteria on the basis of which a service is classified as new are far from clear. Others attack the lack of transparency in deciding on what a new service is in general. Overall, it seems that the definition of a 'new service' is highly dependent on the political context. In that sense, legal certainty is far from enhanced by the Public Value Test. Nevertheless, a clear definition of concepts is key to making credible and legitimate policy.

Balance public value with market impact

A second point that comes to the fore is the necessity of achieving a balance between public and private value. In many sectors, governments pursue public service objectives as well as profit maximisation in order to enhance societal welfare. In every instrument – regardless of the sector it is used in – a balance should be struck between public and private value. It seems that the British Public Value Test is quite successful in doing so[57].

The Public Value Assessment is indeed as critical for the conclusion of the Public Value Test as the Market Impact Assessment. It is apparent that a detailed analysis is the basis of the Public Value Assessment. General aspects like reach of the target audience and inclusiveness have been taken into account in all Public Value Tests so far. Platform neutrality has also been an issue in most tests. The offer of a high definition channel was, for example, dependent on the terrestrial digital television platform Freeview. The Trust insisted and will monitor that

[55] S. Radoslavov and B. Thomass (2010). "The public value test as a societal debate about the future of PSB: The case of the German ZDF". Paper presented at the Ripe Conference 2010 'Public Service Media After The Recessions', London, 9-11 September; K. Donders and C. Pauwels (2010). "The introduction of an ex ante evaluation for new media services: 'Europe' asks it or public broadcasters need it?". *International Journal of Media and Cultural Politics* 6(2), 133-148.

[56] H. Van Den Bulck (2010). "Making Ends Meet: Developing Ex Ante Assessments in Flanders". Presentation at the IES & IBBT-SMIT Conference Exporting The Public Value Test: Assessing EU Member States' evaluation of public broadcasters' new media services, Brussels, 25 May; H. Van Den Bulck, H. and H. Moe (2010). "Public Service Media governance after the crisis: comparing assessment practices in Flanders and Norway". Paper Presented at the RIPE Conference 2010 'Public Media After The Recession', London, 8-11 September; K. Donders and C. Pauwels (2010). "The introduction of an ex ante evaluation for new media services: 'Europe' asks it or public broadcasters need it?". *International Journal of Media and Cultural Politics* 6(2), 133-148.

[57] See also D. Levy (2010). "The British Public Value Test". Presentation at the IES & IBBT-SMIT Conference Exporting The Public Value Test: Assessing EU Member States' evaluation of public broadcasters' new media services, Brussels, 25 May.

the channel will be offered via cable and satellite as well. With regard to seven-day catch-up services, the Trust questioned, among others, the iPlayer's reliance on Windows technology. Citizens making use of older Windows versions, Linux or Apple, were not able to access on-demand services. Again, the Trust encouraged the BBC to overcome these issues, which are mainly due to problems in finding DRM-protection software that meets all demands of the right holding parties the BBC is cooperating with. In contrast with these more general concerns (followed by encouragement and monitoring processes), the Trust has also asked for more specific alterations of services in light of public value concerns. For example, the launch of a Gaelic television channel was considered to provide little public value. The Trust was of the opinion that the educational strategy underlying the Gaelic channel was poor and needed improvement on several levels. Target groups had to be defined better, subtitles were necessary to reach citizens that could not speak the Gaelic language yet (but might improve their knowledge with this channel), marketing plans were not well thought through, etc. The Trust published some provisional conclusions, after which the BBC management adapted its proposal in order to overcome some of the Trust's concerns. Consequently, a final conclusion on the service was published. This process could serve the licence fee payer (forcing the BBC to adapt its proposals in a public value oriented way) and also proves that Public Value Tests are not necessarily predominantly oriented at the assessment of market impact. Obviously, Ofcom's Market Impact Assessments have been important. Yet, public value and the way in which proposed services contributed to fulfilling the BBC's purposes were also at the core of the entire evaluation.

Consensus as a driving force

A third aspect that is apparent relates to the outcome of the process of conducting a Public Value Test and, in addition, the objective underlying it. One can observe that three out of four services were approved, albeit with conditions. This is not a statistically relevant conclusion. However, it shows that Public Value Tests are not oriented at a rejection of services. Conditions relate to public value or market aspects. The imposed changes that have to limit market impact are considerable, but do not necessarily significantly alter a service. For instance, many concrete alterations were suggested by Ofcom and taken over by the BBC Trust in its final conclusions on the BBC's proposal for on-demand services. For example, audio books and classical musical could be offered via the seven-day catch-up service, but were not allowed to be stored by citizens themselves. The latter should prevent harm to those companies selling audio books and classical music. Time limits were also introduced for storage of on-demand content, in order to balance the benefits of the users with the concerns of private competitors. While the BBC originally proposed that licence fee payers could store services for 13 weeks, after reactions of private competitors, the Trust decided to limit this to 30 days. The Trust argued that this was a fair solution to ensure that licence fee payers would be able to access BBC public service content whenever possible, but would not expect other undertakings to offer content for such a wide time frame. The BBC management accepted this condition without criticism. It was also emphasised that the content offered via iPlayer had to be available for other parties and that market conform behaviour should guide the BBC in its relations with these parties.

Collins concludes on the basis of this first Public Value Test that the Trust has no intention to 'rubber-stamp' the BBC's management proposals. The Trust genuinely looks at citizens' preferences and comes to a conclusion, containing quite some conditions to alter aspects of proposed services, based on a continuous deliberation process[58]. The fact that the Trust continues its monitoring of the BBC's performance and compliance with conditions is a case in point. While a service licence was awarded for the on-demand services of the BBC, that does not give the BBC the liberty to take the service forward as it sees fit.

This search for consensus it very important. The media sector has been characterised by opposition between public and private broadcasters since the sector was liberalised in the 1980s. The Public Value Test attempts to overcome the obviously diverging interests of both parties. It seriously considers private sector objections against specific services of public broadcasters and gives the latter the opportunity to adjust aspects of the service in question. All of this can be regarded as a constructive process in which conflict offers no way out, whereas a consensus does. The realisation of a consensus is, moreover, the result of a highly transparent process in which hidden lobbying and bargaining is – at least in part – replaced by formal and open negotiation processes[59].

Limit political influence

A definite positive aspect of the Public Value Test is the way in which it limits government intervention and, more in particular, political influence in market processes. For example, the required alterations of proposed new media services are not determined by government authorities. This is inherent to the design of the Public Value Test. It needs to be stressed once again, that the BBC Trust is responsible for this test and has taken this task seriously. The conditions that accompany approval are inspired by the Trust's desire to protect the interests of the licence fee payers. The latter, so the Trust also acknowledged by taking into account Ofcom's suggestions, is not served by a public broadcaster that stifles market initiatives[60]. The fact that the Trust oversees the Public Value Test is relevant because it limits government intervention with the public broadcaster's activities while at the same it ensures that the public broadcaster is held accountable. At least from a theoretical point of view, this system finds a balance between the need for editorial independence and a holistic public broadcasting project on the one hand and the must of supervision and limits to what the BBC can do on the other hand. Such a system, albeit far from perfect (there is criticism on the independence of the BBC Trust from the BBC's management), requires professionalisation of regulators and advisory bodies.

Include citizens

Finally, the test is also in practice oriented at the licence fee payer's interests. Not only private sector stakeholders, but also licence fee payers respond to the BBC Trust's consultations. This, so one former policy advisor of the BBC said in an interview, has resulted in a change of focus within the BBC itself. The latter is forced to make a positive case for particular public services. As a consequence thereof, the BBC – within its internal decision making process – became

58 R. Collins (2007). "The BBC and 'public value". *Medien und Kommunikationswissenschaft* 65(2), 164-184.

59 K. Donders and C. Pauwels (2010). "The introduction of an ex ante evaluation for new media services: 'Europe' asks it or public broadcasters need it?". *International Journal of Media and Cultural Politics* 6(2), 133-148.

60 M. Maggiore (2009). "The BBC Public Value Test". Presentation at the Workshop on 'European State aid policy and its impact on Member States' regulation of public service broadcasting', Brussels, 4 June.

more oriented at the licence fee payer's interest and less at government preferences. Collins, although also criticising some aspects of the Public Value Test, agrees and argues that

> "[T]he BBC has put forward what is probably the most fully developed set of reflections on public value and the implementation of a public value based regime of any UK public body. The BBC's proposals have been widely seen as thoughtful, self-critical and innovatory"[61].

These, largely positive, observations should not lead to the conclusion that the Public Value Test is an example to be followed by other governments. The bureaucratic burden of the test is considerable for the BBC and other stakeholders (including the private sector, Ofcom and the BBC Trust). Also the extensive market forecast exercises conducted by Ofcom can be questioned. However, the benefits seem to outweigh the costs for the moment.

Conclusions and recommendations

The British Public Value Test is a specific tool to deal with a sector-specific issue, being the settling of competition conflicts between public broadcasters and private media companies on the one hand and the necessity of legitimising public service media as a policy and public value project on the other hand. It should not be considered a best practice as other countries are only starting to explore the Public Value Test as an instrument – with the exception of Germany in which during the last two years up to 40 tests have been conducted[62]. However, it is not worst practice either. Our analysis shows that the implementation of the test has its benefits. It enhances transparency of decision taking, furthers both public value and economic objectives of media policy and is consensus-based instead of conflict-based. In that sense, it was not a bad idea of former Commissioner for Competition Neelie Kroes to ask Member States to implement a model similar to the British Public Value Test in order to comply with the fair competition rules of the Treaty on the Functioning of the European Union[63].

Can this model also be exported to other sectors where the economic and public domain are intertwined? Of course, this question cannot be conclusively answered. However, it seems at least worthwhile to explore the possibility of having a sort of public value test in other sectors. The idea of balancing a set of pre-defined public value objectives with possible market effects seems most adequate to deal with the complexity of the current economic landscape and intervening policies therein. Key is to limit political intervention in order to allow for evidence-based exercises. This is not only paramount for the Public Value Test in the broadcasting sector, but is essential in order to allow for a rational evaluation of the public value of a specific service. To conclude, the procedure of a Public Value Test has to be clearly spelled out in order to prevent problems of interpretation and to make sure that evaluations can be conducted as quickly as possible as to prevent harm to economic development and public service delivery.

---

[61] R. Collins (2007). "The BBC and 'public value". *Medien und Kommunikationswissenschaft* 65(2), 164-184.

[62] S. Radoslavov and B. Thomass (2010). "The public value test as a societal debate about the future of PSB: The case of the German ZDF". Paper presented at the Ripe Conference 2010 'Public Service Media After the Recessions', London, 9-11 September.

[63] N. Kroes (2008). "The way ahead for the Broadcasting Communication". Speech at the Conference on 'Public service media in the digital age', Strasbourg, 17 July.

Conclusions

Conclusions: Future Legal Challenges for Geographic Data and SDI

Katleen Janssen[1] and Joep Crompvoets[2]

The path towards the successful use, dissemination and sharing of geographic data is still filled with legal challenges and obstacles. These legal difficulties are increasingly acknowledged by practioners and academics, but much work still needs to be done in fully identifying their scope, in assessing their affect on the use of geographic data, and in providing solutions that benefit all stakeholders. The contributions in this book provide a sample of legal challenges that the producers and users of geographic data have to face in their daily actions and interactions on the market. They also show examples of new research trying to tackle these challenges, and existing practices on how law and policy are already being applied in a successful way to promote the availability of geographic data.

The contributions reflect some of the discussions that were held in two workshops organised by the Interdisciplinary Centre for Law and ICT of the KU Leuven in the course of the Spatialist project. Spatialist was a 4 year research project, funded by the Flemish Government Agency for Innovation by Science and Technology (Agentschap voor Innovatie door Wetenschap en Technologie, IWT). Its main objective was to determine the requirements for an operational spatial data infrastructure (SDI) in the Flemish region that is compliant with INSPIRE and is efficient, effective, flexible and feasible. The requirements were developed from a multidisciplinary point of view, including geomatics, sociology, economics, law, and public management. In this way, the project aimed to close the gap between the research on technical aspects of SDIs, already at a mature stage at the start of the project in 2007, and the research on non-technical aspects, which at the time was still in its infancy.

During the course of the project, it was found that particularly the legal aspects of geographic data and SDI had not received much attention and that the research community around this topic was very small. In addition, there were few opportunities for interaction around these legal issues with legal practitioners and policymakers, due to the focus of most workshops, conferences and meetings on technical or institutional questions rather than real legal debate. Therefore, the legal workshops intended to bring legal scholars and legal practitioners together to discuss existing and new legal challenges and to move towards a more mature research field on legal aspects of geographic data and SDI.

The chapters in this book discuss legal problems and challenges, and some possible solutions to meet these challenges. They critically reflect on existing legislation and show where actions are needed to improve the functioning of the SDI and optimise the dissemination of geographic data. Because they highlight not only the arguments for increasing the sharing of geographic data, but also the possible problems arising with this increase (e.g. with regard to privacy or liability), the contributions demonstrate an essential

[1] Interdisciplinary Centre for Law and ICT, KU Leuven, katleen.janssen@law.kuleuven.be.
[2] Public Management Institute, KU Leuven, joep.crompvoets@soc.kuleuven.be.

element of any legal discussion on SDI and geographic data: the search for a balance between all the rights and interests involved, taking into account the value given to these rights and interests by international treaties, supranational legislation and national rules. This entails that while for most disciplines a well-functioning SDI is an SDI in which as much geographic data as possible is shared, this is not necessarily the case from a legal point of view. Keeping in mind the necessity of reaching a balance, legal research does not look for the broadest possible level of data sharing, but rather an 'optimal' level, at which legitimate reasons may limit the availability of geographic data in the SDI. For instance, the necessity of data sharing for policy making or for the growth of transparency in some cases may need to be weighed against the fundamental right of an individual to maintain his or her privacy. This weighing of interests often brings the need for case-by-case decisions, sometimes to the frustration of other disciplines that reason in more absolute terms.

In these conclusions, we try to determine some of the future challenges and priorities with regard to the legal aspects of SDIs and geographic data, based on the main conclusions of the contributions in this book and the discussions in the workshops. We will try to group them according to the division made in this book between policies promoting the availability of geographic data, policies restricting this availability, and policies regulating the relationship between the public and the private sector.

Policies promoting availability

From the chapters in this book, it is clear that, even though much has been achieved in the last few years, access and licensing policies for geographic data still require attention from legal scholars and practitioners. The following main topics for future research arise: interoperability of licensing, standard licensing frameworks, and interaction between legal requirements and technical developments relating to rights management, metadata and other technical aspects of the SDI.

Both the Australian and the Greek experience show that governments and public bodies are starting to see the potential benefits of making their geographic data available to the public for increasing transparency and accountability, economic growth and social interaction. Many other countries are also starting to follow a more permissive approach with regard to the dissemination and use of their geographic data. In many cases this is embedded in the more general development towards open government and open government data, as clearly indicated by Prodromos Tsiavos et al. with regard to Greece. The initiatives grouped under the flagship 'open government data' want to make data held by the public sector available for any type of use, from the belief that this would stimulate transparency and innovation. More and more governments and public bodies are answering this call for data and create portals or websites from which their data can be obtained. The United States and the United Kingdom are at the forefront, but many other countries, both from the developed and developing world, are following their lead. Available data do not only include spending data or parliamentary documents, but often also geographic data.

Not only national policymakers are embracing open data, also international and supranational authorities are developing plans and programmes for increasing data sharing. Bastiaan van Loenen et al. and Prodromos Tsiavos et al. already mention INSPIRE, and Janssen and Vandezande refer to the PSI directive. However, there are many more international initiatives that are trying to increase the availability of public sector data, including geographic data. At the EU level, the Global Monitoring for Environment and Security (GMES) programme aims to make its earth observation data available under an open data policy. In the 2010 regulation on GMES, the European Parliament and Council adopted 'full and open access to information produced by GMES services and data collected through GMES infrastructure, subject to relevant international agreements, security restrictions and licensing conditions, including registration and acceptance of user licences'[3]. In the course of 2012, the European Commission will create a GMES data policy based on this principle. Other initiatives for increasing the availability of data involve marine data[4], transport data[5], cultural heritage[6], and scientific data[7]. All these plans and programmes should be embedded in a general policy of the European Commission to increase the availability of data held by the public sector. This 'open data strategy' will be part of the European Commission's package of measures linked to the amendment of the PSI Directive. It is expected that the proposal for this amendment will include a full obligation for public bodies to make their data available for use outside of the public sector.

Not only the European Union is investing in open data policies. At the global level, the Global Environmental Observation System of Systems (GEOSS) aims to increase the availability of satellite and in situ data for empowering the international community to protect itself against natural and human induced disasters, to respond to climate change, to manage ecosystems, conserve biodiversity, etc[8]. The GEOSS data policy should be based on full and open exchange of data, with minimal use restrictions and at minimum cost[9].

All these different initiatives, action plans, programmes and policies for data sharing have to deal with comparable legal questions. One of the main issues that arise here has already been touched upon by Van Loenen et al.: the interoperability of licensing conditions. Each of the initiatives mentioned in the previous paragraphs includes at least recommendations for the types of licences or the conditions under which the data involved should be made available, and many of them refer to particular licences or licensing frameworks (see for example the Australian initiatives described by Anne Fitzgerald).

While all initiatives favour an open approach with as few restrictions as possible, this still does not necessarily mean that legal interoperability can be reached, in the sense that data coming from different sources may not necessarily be combined due to different conditions that are imposed on the use of the various datasets. For instance, if a particular dataset cannot be published on the Internet, this entails that any product or service using these data in combination with other data, cannot be disseminated online if the original data set can

---

[3] European Parliament and Council (2010). Regulation of 22 September 2010 on the European Earth monitoring programme (GMES) and its initial operations (2011 to 2013), *OJ L* 276, 1.

[4] European Commission (2010). Communication to the European Parliament and the Council. Marine Knowledge 2020: Marine data and observation for smart and sustainable growth, COM (2010) 461 final.

[5] European Commission (2008). Communication. Action Plan for the Deployment of Intelligent Transport Systems in Europe, COM (2008) 886 final.

[6] See http://www.europeana.eu/portal/ (accessed on 20/10/2011).

[7] European Commission (2007). Communication to the European Parliament, the Council and the European Economic and Social Committee. on scientific information in the digital age: access, dissemination and preservation, COM (2007) 56 final.

[8] See http://www.earthobservations.org/geoss.shtml (accessed on 20/10/2011).

[9] Group on Earth Observations (2010). *GEOSS Data Sharing Action Plan,* http://www.earthobservations.org/documents/geo_vii/07_GEOSS%20Data%20Sharing%20Action%20Plan%20Rev2.pdf (accessed on 20/10/2011).

be accessed from the product or service, either directly or via 'reverse engineering'[10]. Even 'open licences' can suffer from a lack of interoperability. Usually, a licence is considered open when it contains no restrictions on the use that can be made of the data it covers, but it is still accepted that attribution is required, or that the user is forbidden to misuse the data. Even though in theory all kinds of products and services could be developed combining several datasets, if each of these datasets requires a different attribution clause, the end result of the product risks becoming very user-unfriendly. One can only imagine the problems when less open licences have to be combined. This will effectively render impossible the combination of particular types of data into a new product or service.

While this problem is increasingly being acknowledged by the SDI community, solutions are only emerging very slowly. As Van Loenen et al. describe, the main focus of policy makers has been on the creation of national or sector-based licensing policies, without attention for how these policies should relate or 'inter-operate'. These policies are based on the specific needs of the data providers or use terminology prescribed by national legislation, restricting the possibilities for data users to create cross-provider or cross-border information products or services. More research is needed into how these licensing frameworks can work together and what efforts are still needed to create full legal interoperability. It should be examined in how far the use of national terminology hinders the validity of licences in other jurisdictions and how legal certainty with regard to applicable terms and conditions can be increased.

In addition, the relationship between legal interoperability and the technical developments with regard to rights management, metadata and other technical aspects of the SDI needs to be defined in order to fully guarantee the automatic availability of geographic data via portals and services without the need for human intervention. If technical solutions for rights management and automatically concluding licences are developed separately from legal solutions, the former may end up being problematic to implement in national jurisdictions and the latter may be unrealistic or unadjusted to technical developments and practice. Any future legal research on the interoperability of licences and possible standard licensing frameworks should be conducted in close cooperation with experts on the technical aspects of rights management.

Policies restricting availability

The chapters of the second part of this book deal with legislation and policies that restrict the availability of geographic data. Each of these contributions raises questions that will need further research to guarantee that future SDIs find an appropriate balance between the rights and interests involved in the sharing of geographic data. Relating to privacy, the main challenges stem from the concept of personal data: what is the status of the geographic component of such data; how is the scope of personal data influenced by technological developments; and how can the objectives of an SDI be reconciled with the protection of personal data? With regard to liability and national security, the main question lies in aligning the perception and fear of these possible threats to the availability of geographic data with the actual rules and limitations that may arise.

[10] The debate when a particular information product or service qualifies as a derived product that does or does not include the original and is therefore subject to permission from the original rightholder for further dissemination, is not addressed here.

With regard to privacy, both Pomfret's contribution and Kosta & Gasson focus on location data and the tracking of persons via their connections to a communication network, e.g. GPS, Bluetooth, or remote sensing. Pomfret shows that in the United States there is no encompassing legal framework to deal with location privacy, and that the true scope and impact of the concept of location privacy is still unknown – and very much dependent on the reasonable expectation of privacy, which in itself is an evolving concept, particularly dependent on the continuous developments of technology in this field. Kosta & Gasson demonstrate that tracking the location of a person for a longer period of time can lead to a detailed profile of that person and an invasion of their privacy. They discuss the European legal framework for the collection of location data and argue that this legal framework creates an artificial distinction between location based services that are provided via a public communications network and those that are not. This causes uncertainty for the providers of the location based services and in turn also for the users.

The debate on location privacy is an important one, particularly considering the increasing interest of government and the private sector in crowdsourcing information from the citizens. While crowdsourcing often relies on consent (e.g. because the citizen has sent information on a particular location on his or her own initiative), or on anonymous data, some questions still remain. How far should the consent of the citizen reach: is it sufficient if it is given only once, or should it be repeatedly requested, and if so, at which intervals? If the crowdsourcing data are collected anonymously, the legislation on the processing of personal data does not apply, so there is no need to obtain consent from the data subject. One would also assume that in this case there is no concern for the citizen's privacy, but the continuous development of technology has dramatically increased the potential for de-anonymization of data via the method of so-called 'jigsaw identification', combining information from different databases to build a profile on the basis of which to identify a person[11]. This puts into question the scope of the legislation on the protection of personal data: if data is never really anonymous anymore, do the obligations with regard to the processing of personal data apply to any collection of data, whether anonymous or not? It should be examined how the policymaker, the courts, the companies and the citizens should deal with this.

Not only geographic data traditionally considered as location data (data obtained by tracking people via their communication devices) can cause privacy issues, but also other geographic data, such as cadastral parcels, addresses, mobile mapping imagery, etc. Many of these data can indeed be linked to a particular natural person, qualifying them as personal data. This entails that all the obligations of the legislation on the processing of personal data will apply. In the European Union, this means for instance that the data can only be collected for a specific and justified purpose and that they can only be used further for purposes that are compatible with the original purpose they were collected for. In addition, they cannot be kept longer than necessary for the fulfillment of the purpose they were processed for[12]. These principles can be problematic in an SDI, where the objective is to collect data only once and re-use it for multiple purposes (including future purposes that might currently still be unknown). Privacy authorities around the world struggle with reconciling the needs of government bodies for sharing geographic data with the protection of the privacy of

[11] K. O'Hara (2011). *Transparent Government, Not Transparent Citizens: A Report on Privacy and Transparency for the Cabinet Office*, http://eprints.ecs.soton.ac.uk/22769/ (accessed on 20/10/2011).

[12] See Art. 6 of Directive 95/46/EC of the European Parliament and of the Council of 24 October 1995 on the protection of individuals with regard to the processing of personal data and on the free movement of such data, *OJ L* 281, 31.

the citizen, leading to uncertainty among SDI stakeholders as to what extent geographic personal data can be shared and re-used. More research is needed into how the geographic component of the data involved can be separated from the personal component while still maintaining the fitness-for-purpose of the data, or how the citizen's privacy can be guaranteed while sharing the geographic data.

Cho's contribution focuses on liability issues for geographic data, and highlights a number of technological and societal developments that cause new challenges for traditional liability principles, including GPS navigation and crowdsourcing or volunteered geographic information (VGI). Liability is traditionally based on the fact that a particular person has made an error and caused damage to another person or good, either within the context of a contractual relationship, or outside any contractual commitments. The current services and products based on geographic data can sometimes cause difficulty in applying these general liability principles, because many different data stemming from many different suppliers may be combined with software and hardware again created by multiple vendors. In such situations, it is nearly impossible to determine which party is responsible for any possible errors in the service or product and to whom the damage should be attributed. In the case of VGI, this problem is supplemented by the fact that the data providers are no longer professionals, but rather amateurs that may or may not have had training in the field of the collection or representation of geographic data.

Even though possible liability concerns are often raised as objections – or at least caveats – to the dissemination of geographic data in an SDI, the number of court cases dealing with liability for geographic data, products or services is limited. This might indicate that liability issues rarely arise in the SDI or in the market for geographic information products and services, but it could also be that the liability questions rarely make it to court due to the complicated nature of the geographic information services or products. It should be examined whether liability should actually be considered a limiting factor to the dissemination of geographic data, and why there is so much fear of liability, yet so few liability court cases. More insight is needed in how to apply liability principles to geographic data, services and products, and if the existing liability rules are still suitable to deal with the challenges for disseminating geographic data in an SDI.

Finally, Adhikari discusses the possible threats to national security caused by the dissemination of high resolution satellite imagery. He examines the possibilities for remedying such treats under international space law, and questions the validity of India's policy to limit access to these high resolution images. In the debate on potential reasons to limit access to geographic data, national security is often mentioned by public sector bodies as an argument to restrict access. The counterargument that is usually raised to this, states that comparable data can already be obtained from the private sector, so there is no point in limiting access. The increasing attention for the global sharing of satellite data (e.g. GEOSS, GMES) will only intensify this debate in the future, and a clear understanding should be found as to the extent to which national security can be considered a valid reason for limiting access to geographic data.

Relationship between the public and the private sector

In the third part of the book, attention is given to the relationship between the public and the private sector on the market for geographic data and geographic information products and services. The main priority for future research in this field is finding new structures for cooperation between the sectors, in addition to adopting processes for including the citizens as a third party in this relationship.

Janssen & Vandezande try to set some criteria for determining the role of the public sector in the information market, while Donders & Raats show how the same debate is being tackled in another sector, i.e. public broadcasting. Determining how the public and the private sector should interact remains an issue in all domains where government bodies traditionally have played a big role due to their natural monopolies on e.g. telecom infrastructure, utilities, broadcasting, or the provision of geographic data. Due to technological developments, cultural changes and political decisions, these natural monopolies no longer exist, and the public and private sector should be able to compete on a level playing field in the market. At the same time, the public sector often still has a role to play in providing public services to the citizens. Maintaining a balance between these two roles and adapting to market developments remain challenging for the public sector.

While the conflicts between the public and private sector on the market for geographic information services and products have received considerable attention, the focus should rather be moving to the possible ways in which the public and private sector should cooperate and how public private partnerships can play a role in the development of an SDI. Can creating new ways of cooperation between business and government create new opportunities for SDIs, and how should this cooperation be envisaged in order to guarantee the optimal availability of geographic data and the optimal functioning of the SDI?

The contributions of Donders & Raats and Janssen & Vandezande deal with the relationship between two main players in the SDI, the public sector and the private industry. However, the rise of web 2.0 and user generated content have added a third player to the mix: the citizen. While research in other domains has started to recognise the potential role of VGI, legal research is still in its infancy. Yet, there are many questions to be answered on the role of VGI in the SDI, and how citizens can become a full third pillar of the SDI, next to the public and the private sector. New processes for governance and rights management need to be elaborated for the integration of VGI and datasets traditionally produced by the public sector or industry. In addition, as Cho already mentioned, the division of responsibility, accountability and liability between all the actors involved in a VGI-enabled SDI will create new dilemmas and should be high on the agenda of future research on SDIs and the dissemination of geographic data.

Take legal research on geographic data outside of the box

The previous sections indicate that a large body of research still needs to be developed with regard to the legal aspects of geographic data and SDIs. Next to the topics arising from this book, many other issues also need to be considered. Should a right to data be developed that gives citizens the right to access and use raw data? How should authentic sources be integrated in the daily activities of the SDI stakeholders? How to deal with trust and security concerns in the cloud? These are only a few examples of questions that legal research in an SDI environment will have to tackle in the future.

Whatever legal research is performed, its main prerequisite to have any value for the development of an optimal SDI, is that it is not performed in a vacuum. First, this entails that it should place geographic data and SDI in their broader legal context and aim to advocate legal solutions that provide maximum coherence with the existing legal framework, in order to ensure legal certainty, transparency and user friendliness. In doing so, the research should always question whether the situation or characteristics of geographic data or SDI are so particular that they need a separate approach, or if they fit into a more general framework that applies to a broader spectrum of data or infrastructure. In other words: it should be examined whether – and not assumed that – geographic data and SDI are special or different, requiring special treatment.

Second, for any legal research to lead to successful changes in legislation, policy and practice regarding SDI, it has to be embedded in the technical, cultural and institutional reality of the SDI. As mentioned before, legal provisions that do not allow for evolutions in technology or that cannot be translated into technical requirements need to be avoided. In addition, legal provisions are interpreted and implemented in a cultural and institutional context. These cultural and institutional factors play a big role in the ability of regulatory instruments to have an impact on the organisation and working of an SDI. In many cases, the law may be intended to reach a particular purpose, but its institutional embedding or cultural interpretation may lead to another result. Maintaining continuous interaction with other disciplines can avoid this. For this reason, any legal research on geographic data and SDI should be done in a multidisciplinary context, in close contact with the stakeholders.

This book aimed to show the importance of the legal aspects of disseminating geographic data in an SDI, and it has tried to indicate a number of research priorities for the near future. This would not have been possible without the contributions of all the authors included in this book, but also all the participants of the Spatialist workshops. These workshops have shown that it is vital that the research community gets more opportunities to meet, discuss and work together to tackle the issues that have been mentioned in these conclusions. Therefore, we would like to express our thanks to all of the people who participated or have expressed interest in this project, and encourage all legal researchers and practitioners to increase their knowledge and experience sharing in order to develop a true acquis of legal expertise on geographic data and SDI.

List of Main References

Adhikari, M. (2010). *Legal Perspective of Remote Sensing Data Dissemination for Earth Observation with Special Reference to India*, M.Phil. Dissertation. Hyderabad: NALSAR University of Law

Antoniadis, A. (2006). "The financing of public service broadcasting". In Rydelski, M.S. (ed.). *The EC State aid regime: distortive effects of State aid on competition and trade*. London: Cameron May, 591-630

Article 29 Working Party (2005). *Opinion 5/2005 on the use of location data with a view to providing value added services*, WP115, http://ec.europa.eu/justice/policies/privacy/docs/wpdocs/2005/wp115_en.pdf (accessed on 19/11/2011)

Ashbrook, D. and Starner, T. (2003). "Using GPS to learn significant locations and predict movement across multiple users". *Personal and Ubiquitous Computing* 7(5), 275-286

Aslama, M. and Syvertsen, T. (2007). "Public service broadcasting and new technologies: marginalisation or re-monopolisation". In De Bens, E. (ed.). *Media between culture and commerce*. Bristo: Intellect, 167-179

Athanasiou, S. (2010). *Report on the INSPIRE transposition public consultation*. Athens: MEECC

Atkinson, B. (2007). *The True History of Copyright: The Australian Experience 1905 – 2005*. Sydney: Sydney University Press

Attorney-General's Department (2010). *Statement of IP Principles for Australian Government Agencies*, http://www.ag.gov.au/Documents/Statement%20of%20IP%20Principles%20for%20Australian%20Government%20Agencies2.pdf (accessed on 18/10/2011)

Australian Government (2010). *Government Response to the Report of the Government 2.0 Taskforce*, http://www.finance.gov.au/publications/govresponse20report/doc/Government-Response-to-Gov-2-0-Report.pdf (accessed on 18/10/2011)

Australian Government. Office of Spatial Data Management (2010). *Australian Government Spatial Data Activities 2009: A report of the Spatial Data Management Group*, http://spatial.gov.au/sites/default/files/legacy/osdm.gov.au/SDMG-AR%3d2009c945.pdf?ID=1043 (accessed on 18/10/2011)

Australian Government. Office of the Australian Information Commissioner (2011). *Principles on open public sector information*, http://www.oaic.gov.au/publications/agency_resources/principles_on_psi_short.html (accessed on 18/10/2011)

Australian Land Information Council (ALIC) (1990). *Data Custodianship/Trusteeship, Issues in Land Information Management*. Belconnen, ACT: ALIC

Australian Ocean Data Network (2011). *AODN Data Policy (version 1.0)*, http://imos.org.au/fileadmin/user_upload/shared/AODN/AODN_data_policy_v1.pdf (accessed on 18/10/2011)

Australian Spatial Consortium (2008). *Submission 307 to the Review of the National Innovation System*, http://www.innovation.gov.au/Innovation/Policy/Pages/ReviewoftheNationalInnovationSystem.aspx (accessed on 18/10/2011)

Banerjee, I. and Seneviratne, K. (ed.) (2005). *Public Service Broadcasting: A Best Practices Source Book*. Paris: UNESCO

Bardoel, J. and d'Haenens, L. (2008). "Public service broadcasting in converging media modalities. Practices and reflections from the Netherlands". *Convergence. The International Journal of Research into New Media Technologies* 14(3), 351-360

Bardoel, J. and Ferrell Lowe, G. (2007). "From public service broadcasting to public service media. The core challenge". In Ferrell Lowe, G. and J. Bardoel (ed.). *From public service broadcasting to public service media*. Göteborg: Nordicom, 9-28

Bardoel, J. and Vochteloo, M. (2009). "Media policy between Europe and the nation-state: the case of the EU Broadcast Communication 2009". Paper presented at the ECREA-CLP Workshop on 'new direction for communication policy research'. Zürich, 6-7 November

Barnett, S. (2002). "Which end of the telescope? From market failure to cultural value". In Cowling, J. and Tambini, D. (ed.). *From public service broadcasting to public service communications*. London: IPPR, 34-45

Barnett, S. (2006). "Public service broadcasting: a manifesto for survival in the multimedia age (a case study of the BBC's new charter)". Paper presented at the RIPE Conference, Amsterdam, 2-6 October

Baten, I. and Van der Starre, G. (1996). *Elektronische toegankelijkheid van overheidsinformatie*. The Hague: Rathenau Instituut [in Dutch]

Bavasso, A. (2006). "Chapter 17: Broadcasting". In Hancher, L., Ottervanger, T. and Slot, P.J. (ed.). *EC State Aids*. London:Thomson Sweet and Maxwell, 420-438

BBC (2004). *Building public value: renewing the BBC for a digital world*. http://downloads.bbc.co.uk/aboutthebbc/policies/pdf/bpv.pdf (accessed on 10/08/2011)

BBC Trust (2007). *Gaelic digital services: Public Value Assessment*. http://www.bbc.co.uk/bbctrust/assets/files/pdf/consult/gaelic_digital_service/pva.pdf (accessed on 10/08/2011)

BBC Trust (2007). *Public Value Assessment: BBC HDTV proposal*, www.bbc.co.uk/bbctrust/assets/files/pdf/consult/hdtv/pva.pdf (accessed on 10/08/2011)

BBC Trust (2007). *BBC on-demand proposals: Public Value Test final conclusions*, http://www.bbc.co.uk/bbctrust/assets/files/pdf/consult/decisions/on_demand/decision.pdf (accessed on 10/08/2011)

BBC Trust (2007). *Public Value Test (PVT): Guidance on the conduct of the PVT*, http://www.bbc.co.uk/bbctrust/assets/files/pdf/regulatory_framework/pvt/pvt_guidance.pdf (accessed on 10/08/2011)

BBC Trust (2009). *Local Video: Public Value Test conclusions*, http://www.bbc.co.uk/bbctrust/assets/files/pdf/consult/local_video/decision.pdf (accessed on 10/08/2011)

Beers, A. (1996). "Openbaarheid van overheidsinformatie". In Baten, I. and Van der Starre, G. (ed.). *Elektronische toegankelijkheid van overheidsinformatie*. The Hague: Rathenau Instituut, 51-77 [in Dutch]

Berne Convention (1886). *Convention of 9 September 1886 for the Protection of Literary and Artistic Works*. http://www.wipo.int/treaties/en/ip/berne/ (accessed 24/10/2011)

Brown, A. (1996). "Economics, public service broadcasting, and social values". *The Journal of Media Economics* 9(1), 3-15

Bruguière, J.M. (2002). *Les données publiques et le droit*. Paris: Editions Litec [in French]

Brunsson, N. et al. (2000). *A World of Standards*. Oxford: Oxford University Press

Buckeridge, R. and Cutler, T. (1994). *Commerce in Content: Building Australia's International Future in Interactive Multimedia Markets. Report for the Department of Industry, Science and Technology, CSIRO and the Broadband Services Expert Group*, http://www.nla.gov.au/misc/cutler/cutlercp.html (accessed on 18/10/2011)

CCRS (2008). *Report for Congress. Satellite Surveillance: Domestic Issues*, www.fas.org/sgp/crs/intel/RL34421.pdf (accessed on 23/06/2011)

Center for Democracy and Technology (2006). *Digital Search and Seizure: Updating Privacy Protections to Keep Pace with Technology*, https://www.cdt.org/publications/digital-search-and-seizure.pdf (accessed on 17/10/2011)

Cheung, P. (2009). "Opening statement". Presentation at 18th session of the United Nations regional cartographic conference for Asia and the Pacific, Bangkok, http://unstats.un.org/unsd/methods/cartog/Asia_and_Pacific/18/OpeningStatement-PC.pdf (accessed on 22/10/2010)

Cho, G. (2005). *Geographic Information Science: Mastering the Legal Issues*. Chichester: John Wiley & Sons

Coleman, D., Georgiadou, Y. and Lebonte, J. (2009). "Volunteered Geographic Information: the nature and motivation of produsers". *International Journal of Spatial Data Infrastructures Research* 4, 332-358

Collins, R. (2007). "The BBC and public value". *Medien und Kommunikationswissenschaft* 65(2), 164-184

Collins, R. (2010). "Exporting the Public Value Test: assessing EU Member States' evaluation of public broadcasters' new media services. The United Kingdom". Presentation at the IES & IBBT-SMIT Conference Exporting the Public Value Test: Assessing EU Member States' evaluation of public broadcasters' new media services. Brussels, 25 May

Collins, R. et al. (2001). "Public service broadcasting beyond 2000: Is there a Future for Public Service Broadcasting?". *Canadian Journal of Communication* 26(1), 3-15, http://www.cjc-online.ca/viewarticle.php?id=615 (accessed on 10/08/2011)

Commonwealth Interdepartmental Committee on Spatial Data Access and Pricing (2001). *A Proposal for a Commonwealth Spatial Data Access and Pricing Policy*, http://spatial.gov.au/sites/default/files/legacy/osdm.gov.au/OSDM/Policies%2band%2bGuidelines/Spatial%2bData%2bAccess%2band%2bPricing/default.html (accessed on 22/5/2009)

Communities and Local Government Committee (2008). *Ordnance Survey. Fifth Report of Session 2007-08*, http://www.publications.parliament.uk/pa/cm200708/cmselect/cmcomloc/268/26802.htm (accessed on 17/04/2011)

Cook, J. (2009). *Economic issues in funding and supplying public sector information*, http://eprints.qut.edu.au/27832/ (accessed on 18/10/2011)

Cook, J. (2010). *A summary view of government cost recovery policies in Australia and New Zealand relating to the supply of public sector information*, http://eprints.qut.edu.au/31609/ (accessed on 18/10/2011)

Copyright Law Review Committee (2005). *Crown Copyright*, http://www.ag.gov.au/Copyright/CopyrightLawReviewCommittee/Reports/Crowncopyright/Pages/default.aspx (accessed on 18/10/2011)

Council of Europe (1950). *Convention for the Protection of Human Rights and Fundamental Freedoms*. http://conventions.coe.int/Treaty/en/Treaties/html/005.htm (accessed 24/10/2011)

Council (1985). Directive 85/374/EEC on the approximation of the laws, regulations and administrative provisions of the Member States concerning liability for defective products, *OJ L* 210, 29

Council (1989). Directive 89/552/EEC of 3 October 1989 on the coordination of certain provisions laid down by Law, Regulation or Administrative Action in Member States concerning the pursuit of television broadcasting activities, *OJ L* 298, 23

Council (2003). Directive 93/13/EEC of 5 April 1993 on Unfair Terms in Consumer Contracts, *OJ L* 95, 29

Council of the European Union (2009). Common Position (EC) No 16/2009 adopted by the Council on 16 February 2009 with a view to the adoption of a Directive 2009/…/EC of the European Parliament and of the Council of … amending Directive 2002/22/EC on universal service and users' rights relating to electronic communications networks, Directive 2002/58/EC concerning the processing of personal data and the protection of privacy in the electronic communications sector and Regulation (EC) No 2006/2004 on consumer protection cooperation protection laws, *OJ C* 103, E/40

Crompvoets, J., et al. (2008). "Interdisciplinary Research Project: SPATIALIST; Spatial Data Infrastructures and Public Sector Innovation in Flanders (Belgium)". *Proceedings of GSDI-10 Conference, Small Island Perspectives on Global Challenges: The Role of Spatial Data in Supporting a Sustainable Future*. St. Augustine, Trinidad

Cuijpers, C. and Koops, B.J (2008). "How fragmentation in European law undermines consumer protection: the case of location-based services". *European Law Review* 33, 880-897

Cuijpers, C. and Pekárek, M. (2011). "The regulation of Location Based Services: Challenges to the European Union Data Protection Regime". *Journal of Location Based Services* 5(4), 223-241

Cunningham, S. et al. (2007). "Why Governments and Public Institutions Need to Understand Open Content Licensing". In Fitzgerald, B., Coates, J. and Lewis, S. (ed.). *Open Content Licensing: Cultivating the Creative Commons*. Sydney: Sydney University Press, 74-92

Cutler & Company (2008). *Venturous Australia - Building Strength in Innovation. Report on the Review of the National Innovation System, prepared for the Australian Government Department of Innovation, Industry, Science and Research*, http://www.innovation.gov.au/Innovation/Policy/Pages/ReviewoftheNationalInnovationSystem.aspx (accessed on 18/10/2011)

Das, S.K. (2007). *Touching Lives- The Little Known Triumphs of the Indian Space Programme. 1st ed.* New York: Black Classics

David, P.A. (1985). "Clio and the economics of QWERTY". *American Economic Review* 75(2), 332-336

Davies, G. (2008). "Methodology for Market Impact Assessments". Presentation at the OFCOM Information Seminar 'Market Impact Assessment of the BBC's new on-demand proposals'. London, 18 September

Department of Broadband, Communications and the Digital Economy (2009). *Australia's Digital Economy: Future Directions*, http://www.dbcde.gov.au/digital_economy/what_is_the_digital_economy/ australias_digital_economy_future_directions/final_report (accessed on 18/10/2011)

Department of Innovation, Industry, Science and Research (2009). *Powering Ideas: An Innovation Agenda for the 21st Century*, http://www.innovation.gov.au/innovation/policy/pages/PoweringIdeas.aspx (accessed on 18/10/2011)

Department of the Environment (1987). *Handling geographic information. Report of the Committee of Enquiry chaired by Lord Chorley*. London: HMSO

Department of Trade and Industry (2005). *The empirical economics of standards*, www.bis.gov.uk/files/file9655.pdf (accessed on 26/08/2011)

DiBiase, D. (2008). "Scoping GIS for education: Making sense of academic and practitioner perspectives", *Geography Compass* 2(5), 1506-1528, http://dusk.geo.orst.edu/ethics/cases/MAPPS.case.pdf (accessed 17/10/2011)

Digital Content Industry Strategic Industry Leaders Group (2006). *Unlocking the Potential: Digital Content Industry Action Agenda Report*, http://www.archive.dcita.gov.au/2007/12/unlocking_the_potential_digital_content_industry_action_agenda_report (accessed on 18/10/2011)

Donders, K. (2010). "Europese Staatssteunregels en de openbare omroep: acties van de Europese Commissie en de Europese gerechtshoven tussen 1992 en 2009". *Mediaforum* 9, 270-278 [in Dutch]

Donders, K. and Pauwels, C. (2008). "Does EU policy challenge the digital future of public service broadcasting? An analysis of the Commission's State aid approach to digitization and the public service remit of public broadcasting organizations". *Convergence. The International Journal of Research into New Media Technologies* 14(3), 295-311

Donders, K. and Pauwels, C. (2009). "European State aid rules and the public service broadcasting remit in the digital age: analyzing a contentious part of European policy and integration". In

Garcia-Blanco, I., Van Bauwel, S. and Cammaerts, B. (ed.). *Media agoras: democracy, diversity and communication*. Cambridge: Cambridge Scholars Publishing, 178-197

Donders, K. and Pauwels, C. (2010). "The introduction of an ex ante evaluation for new media services: 'Europe' asks it or public broadcasters need it?". *International Journal of Media and Cultural Politics* 6(2), 133-148

Dulong de Rosnay, M. (2010). *Creative Commons Licenses Legal Pitfalls: Incompatibilities and Solutions*, http://www.creativecommons.nl/downloads/101220cc_incompatibilityfinal.pdf (accessed on 26/08/2011)

Dusollier, S. (2006). "The Master's Tools v. The Master's House: Creative Commons v. Copyright". *Columbia Journal of Law & Arts* 29, 271-293

Eckhart, J. (2008) "Zehnter Teil. Telekommunikationsgesetz". In Spindler, G. and Schuster, F. (ed.). *Recht der elektronischen Medien*. Munich: Verlag C.H. Beck, 809-1400 [in German]

Elkin-Koren, N. (2006). "Exploring Creative Commons: A Skeptical View of a Worthy Pursuit". In Hugenholtz, B. and Guibault, L. (ed.). *The Future of the Public Domain*. Alphen a/d Rijn: Kluwer Law International, http://ssrn.com/paper=885466 (accessed on 24/10/2011)

Elwood, S. (2008). "Volunteered geographic information: future research directions motivated by critical, participatory, and feminist GIS". *GeoJournal* 72(3-4), 173-183

Epstein, E.F. (1991). "Legal aspects of geographic information systems". In Maguire, D.J., Goodchild, M.F. and Rhind D.W. (ed.) *Geographic Information Systems: Principles and Applications*. London: Longman, 489-502

European Commission (2001). Communication to the European Parliament, the Council, the European Economic and Social Committee and the Committee of the Regions. eEurope 2002: Creating an EU framework for the exploitation of public sector information. COM (2001) 607 final

European Commission (2001). Communication on the application of State aid rules to public service broadcasting, *OJ C* 320, 5

European Commission (2002). Proposal for a Directive of the European Parliament and of the Council on the re-use and commercial exploitation of public sector documents, COM (2002) 207 final

European Commission (2007). Communication to the European Parliament, the Council and the European Economic and Social Committee. On scientific information in the digital age: access, dissemination and preservation, COM (2007) 56 final

European Commission (2007). Meeting on Implementation Issues surrounding the Data Retention Directive (Directive 2006/24/EC). Meeting report, http://www.cullen-international.com/cullen/multi/doc/security/meeting140307_data_retention.pdf (accessed on 11/10/2011)

European Commission (2008). Communication. Action Plan for the Deployment of Intelligent Transport Systems in Europe, COM (2008) 886 final

European Commission (2009). Communication on the application of the State aid rules to public service broadcasting, *OJ C* 257, 1

European Commission (2009). Communication to the European Parliament, the Council, the European Economic and Social Committee and the Committee of the Regions. Re-use of Public Sector Information – Review of directive 2003/98/EC, COM (2009) 212 final

European Commission (2011). Results of the online consultation of stakeholders. Review of the PSI Directive, http://ec.europa.eu/information_society/policy/psi/docs/consultations/cons2010/Results%20of%20the%20online%20consultation%20of%20stakeholders%20final.doc (accessed on 15/04/2011)

European Commission (2011). Overview of the responses to the consultation on Directive 2003/98/EC on the re-use of public sector information, http://ec.europa.eu/information_society/policy/psi/docs/consultations/cons2010/responses.xls (accessed on 18/02/2011)

European Parliament and Council (1995). Directive 95/46/EC of 24 October 1995 on the processing on the protection of individuals with regard to the processing of personal data and on the free movement of such data, *OJ L* 281, 31

European Parliament and Council (1996). Directive 96/9/EC of 11 March 1996 on the legal protection of databases, *OJ L* 77, 20

European Parliament and Council (2002). Directive 2002/58/EC of 12 July 2002 concerning the processing of personal data and the protection of privacy in the electronic communications sector *OJ L* 201, 37

European Parliament and Council (2003). Directive 2003/4/EC of 28 January 2003 on public access to environmental information and repealing Council Directive 90/313/EEC, *OJ L* 41, 26

European Parliament and Council (2003). Directive 2003/98/EC of 17 November 2003 on the re-use of public sector information, *OJ L* 245, 90

European Parliament and Council (2006). Directive 2006/24/EC of 15 March 2006 on the retention of data generated or processed in connection with the provision of publicly available electronic communications services or of public communications networks and amending Directive 2002/58/EC, *OJ L* 105, 54

European Parliament and Council (2007). Directive 2007/2/EC of 14 March 2007 establishing an Infrastructure for Spatial Information in the European Community (INSPIRE), *OJ L* 108, 1

Farchy, J. (2009). "Are Free Licenses Suitable For Cultural Works?". *European Intellectual Property Review* 31(5), 255-263

Feinaeugle, C.A. (2007). "The UN Principles on Remote Sensing and The GATS: Conflicts or Peaceful Co-existence?". *Proceedings of the 50th International Colloquium of Law of Outer Space*. Washington: American Institute of Aeronautics & Astronautics, http://www.iafastro.net/iac/archive/browse/IAC-07/E6/4/7237/ (accessed on 11/10/2011)

Ferrell Lowe, G. (2010). "Beyond altruism: why public participation in public service media matters". In Ferrell Lowe, G. (ed.). *The public in public service media*. Göteborg: Nordicom, 9-36

Fitzgerald, A. (2009). *Open Access Policies, Practices and Licensing: A review of the literature in Australia and selected jurisdictions*, http://eprints.qut.edu.au/28026/ (accessed on 18/10/2011)

Fitzgerald, A. (2010). "Open access and public sector information: policy developments in Australia and key jurisdictions". In B. Fitzgerald (ed.). *Access to Public Sector Information: Law, Technology and Policy*. Sydney: Sydney University Press, http://eprints.qut.edu.au/31024/ (accessed on 18/10/2011)

Fitzgerald, A. (2010). "Legal Issues and Experiences from an Australia Perspective". Presentation at GSDI-12 Conference 'Realising Spatially Enabled Societies'. Singapore, 19-22 October

Fitzgerald, A., Fitzgerald, B. and Hooper, N. (2010). "Enabling open access to public sector information with Creative Commons Licences: the Australian experience". In Fitzgerald, B. (ed.). *Access to Public Sector Information: Law, Technology and Policy*. Sydney: Sydney University Press, http://eprints.qut.edu.au/29773/ (accessed on 18/10/2011)

Fitzgerald, A. and Pappalardo, K. (2009). *Report to the Government 2.0 Taskforce: Project 4 – Copyright Law and Intellectual Property*, http://eprints.qut.edu.au/29416/ (accessed on 18/10/2011)

Fitzgerald, B. (2006). "Open Content Licensing (OCL) for Open Educational Resources". Paper presented at the OECD Expert Meeting on Open Educational Resources, Malmo, 6 - 7 February

Fitzgerald, B. (ed.) (2010). *Access to Public Sector Information: Law, Technology and Policy*, Sydney: University of Sydney Press, http://eprints.qut.edu.au/34085/ (accessed on 18/10/2011)

Fitzgerald, B. et al. (2006). *Creating a Legal Framework for Copyright Management of Open Access within the Australia Academic and Research Sector*, www.oaklaw.qut.edu.au/files/LawReport/OAK_Law_Report_v1.pdf (accessed on 18/10/2011)

Fitzgerald, B. et al. (2011). *Internet and E-Commerce Law, Business and Policy.* Sydney: Thomson Reuters

Fomtchenko, M.M. and Movlyav, A.S. (1997). "High Resolution Remote Sensing: New Aspects and Problems". *Proceedings of the Colloquium of Outer Space*. Washington: American Institute of Aeronautics & Astronautics, 85- 93

Fornefeld, M. et al. (2009). *Assessment of the Re-use of Public Sector Information (PSI) in the Geographical Information, Meteorological Information and Legal Information Sectors*, http://www.epsiplatform.eu/media/files/micus_studie_psi_eu_long (accessed on 20/02/2011)

Franklin, B. (2005). "Introduction". In Franklin, B. (ed.). *Television policy: the MacTaggart Lectures.* Edinburgh: Edinburgh University Press, 1-25

Gadzheva, M. (2007). "Privacy concerns pertaining to location-based services". *International Journal of Intercultural Information Management* 1(1), 49-57

Garnham, N. (1990). *Capitalism and communication: global culture and the economics of information.* London: Sage Publications

Garretti, L. et al. (2009). "Standard Licences for Geographic Information: the Development and Implementation in Local Government in Italy". In van Loenen, B., Zevenbergen, J.A. and Besemer, J.W.J. (ed.). *SDI Convergence - Research, Emerging Trends, and Critical Assessment.* Delft: NCG, 53-62

Gibbons, T. (2005). "Competition policy and regulatory style – issues for OFCOM". *Info* 7(5), 42-51

Gleeson, D. (2007). "The governance of the new BBC". Presentation to the BBC International Day. London, 7 September

Goodchild, M. (2009). "Neography and the nature of geographic expertise". *Journal of Location Based Services* 3(2), 82-96

Government 2.0 Taskforce (2009). *Engage: Getting on with Government 2.0*, http://www.finance.gov.au/publications/gov20taskforcereport/doc/Government20TaskforceReport.pdf (accessed on 18/10/2011)

Groot, N. et al. (2007). "Toegankelijkheid publieke geodata vooral gehinderd door juridische barrieres". *VIMatrix* 15(8), 6-8 [in Dutch]

Group on Earth Observations (2010). *GEOSS Data Sharing Action Plan*, http://www.earthobservations.org/documents/geo_vii/07_GEOSS%20Data%20Sharing%20Action%20Plan%20Rev2.pdf (accessed on 18/02/2011)

Grus, L., Crompvoets, J. and Bregt, A. (2007). "Multi-view SDI assessment Framework". *International Journal of Spatial Data Infrastructures Research* 2, 33-53

Hashimoto, Y. (2003). "Remote Sensing Satellite and Promotion of Regional Security". *Proceedings of the 46th International Colloquium of Outer Space*. Washington: American Institute of Aeronautics & Astronautics, 234-238

Hladjk, J. (2009). "Location Based Services: European data protection rules for mobile commerce". *Privacy and Security Law Report* 8, 21-26

HEMCO (2008). *Inventory of Map Archives of Agencies and Bodies of the Public Sector*. Athens: HEMCO

HEMCO (2009). *National Geographical Data Infrastructure Technical Report*. Athens: HEMCO

Hughes, T.P. (1987). "The evolution of large technological systems". In Bijker, W. et al. (ed.). *The Social Construction of Technological Systems.* Cambridge: MIT Press, 51-82

Information Security and Privacy Advisory Board (2009). *Towards a 21st Century Framework for Federal Government Privacy Policy*, http://www.securityprivacyandthelaw.com/uploads/file/ispab-report-may2009.pdf (accessed on 23/06/2011)

INSPIRE Drafting Team on Data and Service Sharing (2010). *Good practice in data and service sharing*, http://inspire.jrc.ec.europa.eu/documents/Data_and_Service_Sharing/GoodPractice_%20DataService%20Sharing_v1.1.pdf (accessed on 26/08/2011)

INSPIRE Drafting Team Data and Service Sharing (2010). *Guidance on the 'Regulation on access to spatial data sets and services of the Member States by Community institutions and bodies under harmonised conditions*, http://inspire.jrc.ec.europa.eu/documents/Data_and_Service_Sharing/DSSDraftGuidancedocument_v4.1.pdf (accessed on 26/08/2011)

Jakubowicz, K. (2007). *Public Service Broadcasting: a New Beginning, or the Beginning of the End*, http://www.coe.int/t/dghl/standardsetting/media/Doc/PSB_Anewbeginning_KJ_en.pdf (accessed on 10/02/2008)

Janssen, K. (2008). "A legal approach to assessing Spatial Data Infrastructures". In Crompvoets, J. et al. (ed.). *A Multi-View Framework to Assess Spatial Data Infrastructures*. Melbourne: University of Melbourne, 255-272

Janssen, K. (2010). *The Availability of Spatial and Environmental Data in the European Union. At the Crossroads between Public and Economic Interests*. Alphen a/d Rijn: Kluwer Law International

Janssen, K. and Dumortier, J. (2007). "The protection of maps and spatial databases in Europe and the United States by Copyright and the Sui Generis Right". *John Marshall Journal of Computer and Information Law* 24, 195-225

Janssen, K. and Dumortier, J. (2007). "Legal Framework for a European Union Spatial Data Infrastructure: Uncrossing the Wires". In Onsrud, H. (ed.). *Research and Theory in Advancing Spatial Data Infrastructure Concepts*. Redlands: ESRI Press, 231-244

Janssen, K., Van Loenen, B. and Remetey-Fulopp, G. (2011). "Harmonising Licences on a Global Level: Mission Impossible or Piece of Cake?". *Proceedings of the INSPIRE conference, Edinburgh*, http://inspire.jrc.ec.europa.eu/events/conferences/inspire_2011/abstracts/240.doc (accessed on 26/08/2011)

Jones, A. and Sufrin B. (2008). *EC Competition Law: text, cases and materials. 3rd ed.* Oxford: Oxford University Press

Joseph, G. (2008). *Fundamentals of Remote Sensing. 2nd ed.* Hyderabad: Universities Press

Kabel, J. et al. (2001). *Kennisinstellingen en informatiebeleid. Lusten en lasten van de publieke taak.* Amsterdam: Universiteit van Amsterdam [in Dutch]

Kamps, M. (2004). "Thunder and Lightning: Public Sector Information Policy Experiences of Private Meteorological Service Providers". In Aichholzer, G. and Burkert, H. (ed.). *Public Sector Information in the Digital Age. Between Markets, Public Management and Citizens' Rights.* Cheltenham: Edward Elgar Publishing Ltd, 163-186

Kerr, I. and McGill, J. (2007). "Emanations, Snoop Dogs and Reasonable Expectation of Privacy". *Criminal Law Quarterly* 52(3), 393-432

Kosta, E. et al. (2008). "Legal considerations on privacy-enhancing Location Based Services using PRIME technology". *Computer Law and Security Review* 24(2), 139-146

KPMG (2007). *Ministerie van Waterstaat. Evaluatie van de wet op het KNMI*, http://www.knmi.nl/cms/content/8524/wet_op_het_knmi_geevalueerd (accessed on 16/04/2011) [in Dutch]

Krechmer, K. (2009). "Open standards: A Call for Change". *IEEE Communications Magazine* (May), 88-94

Kroes, N. (2008). "The way ahead for the Broadcasting Communication". Speech at the Conference on 'Public service media in the digital age'. Strasbourg, 17 July, http://europa.eu/rapid/pressReleasesAction.do?reference=SPEECH/08/396&format=HTML&aged=0&language=EN&guiLanguage=en (accessed on 18/10/2011)

Lee-Wright, P. (2008). "Virtual News: BBC News at a 'future media and technology crossroads'". *Convergence. The International Journal of Research into New Media Technologies* 14(3), 249-260

Lerner, J. and Tirole, J. (2005). "The scope of open source licensing". *Journal of Law, Economics and Organisation* 21(1), 20-56

Leurdijk, A. (2007). "Public service media dilemmas and regulation in a converging landscape." In Ferrell Lowe, G. and Bardoel, J. (ed.). *From public service broadcasting to public service media.* Göteborg: Nordicom, 71-86

Levy, D. (2010). "The British Public Value Test". Presentation at the IES & IBBT-SMIT Conference Exporting The Public Value Test: Assessing EU Member States' evaluation of public broadcasters' new media services, Brussels, 25 May

Leys, C. (2001). *Market-driven politics: neoliberal democracy and the public interest.* London: Verso

Library of Australia (1994). *Submission to Australian Science & Technology Council Inquiry into Research Data Networks,* http://trove.nla.gov.au/work/13936350 (accessed on 18/10/2011)

Looms, P.O. (2006). "Public service media: all things to all people – on all platform, anytime?". In Nissen, C.S. (ed.). *Making a difference: public service broadcasting in the European landscape.* Eistleigh: John Libbey Publishing, 95-114

Maggiore, M. (2009). "The BBC Public Value Test". Presentation at the Workshop on 'European State aid policy and its impact on Member States' regulation of public service broadcasting'. Brussels, 4 June

Mak, C. (2008). *Fundamental rights in European contract law: A comparison of the impact of fundamental rights on contractual relationships in Germany, the Netherlands, Italy, and England.* Alphen a/d Rijn: Kluwer Law International

Masser, I. (2010). *Building European Spatial Data Infrastructures.* Redlands: ESRI Press

Meixner, H. and Frank, A. (1997). *Study on Policy Issues Relating to Geographic Information in Europe,* ftp://ftp.geoinfo.tuwien.ac.at/frank/2184_GIPolicy.pdf (accessed on 26/08/2011)

Michalis, M. (2010). "The British Public Value Test: Structure and implementation". Presentation at the IES & IBBT-SMIT Conference Exporting The Public Value Test: Assessing EU Member States' evaluation of public broadcasters' new media services. Brussels, 25 May

MICUS (2008). *Assessment of the re-use of public sector information (PSI) in the geographic information, meteorological information and legal information sector. Final report,* http://ec.europa.eu/information_society/policy/psi/docs/pdfs/micus_report_december2008.pdf (accessed on 26/08/2011)

Moe, H. (2008). "Dissemination and dialogue in the public sphere: a case for Public Service Media Online". *Media, Culture & Society* 30(3), 319-336

Murdock, G. (2004). "Building the digital commons: public broadcasting in the age of the Internet". Paper presented at the 2004 Spry Memorial Lecture. Montreal, 22 November, https://pantherfile.uwm.edu/type/www/116/Theory_OtherTexts/Theory/Murdock_BuildingDigitalCommons.pdf (accessed on 18/10/2011)

Murdock, M. (2005). "Building the digital commons. Public broadcasting in the Age of the Internet". In Lowe, G.F. and Jauert, P. (ed.). *Cultural Dilemmas in Public Service Broadcasting.* Göteborg: Nordicom, 213-229

National Research Council (1993). *Toward a coordinated spatial data infrastructure for the nation.* Washington: National Academies Press

National Research Council (2004). *Licensing Geographic Data and Services.* Washington: National Academies Press

Nedović-Budić, Z., Crompvoets, J. and Georgiadou, Y. (ed.) (2011). *Spatial Data Infrastructures in Context: North and South.* Boca Raton: CRC Press

Nitsche, I. (2001). *Broadcasting in the European Union: the Role of Public Interest in Competition Analysis.* The Hague: TMC Asser Press

Nouel, B. (1996). "Commercialisation des données publiques et concurrence". In European Commission (ed.). *Proceedings of the European Commission Conference: Access to Public*

Information. A Key to Commercial Growth and Electronic Democracy. Stockholm: European Commission, 1-9 [in French]

Obermeyer, N.J. and Pinto, J.K. (2008). *Managing Geographic Information Systems. 2nd ed.* New York: The Guildford Press

OECD (2008). *Shaping Policies for the Future of the Internet Economy*, http://www.oecd.org/dataoecd/1/28/40821729.pdf (accessed on 18/10/2011)

OECD (2008). *Recommendation of the Council for Enhanced Access and More Effective Use of Public Sector Information*, http://www.oecd.org/dataoecd/0/27/40826024.pdf (accessed on 18/10/2011)

OECD (2008). *The Seoul Declaration for the Future of the Internet Economy*, http://www.oecd.org/dataoecd/49/28/40839436.pdf (accessed on 18/10/2011)

OFCOM (2004). *OFCOM review of public service broadcasting: Phase 1 – Is television special?*, http://stakeholders.ofcom.org.uk/consultations/psb/ (accessed on 10/08/2011)

OFCOM (2007). *BBC new on-demand proposals – Market Impact Assessment: executive summary*, http://stakeholders.ofcom.org.uk/market-data-research/tv-research/bbc-mias/ondemand/bbc-ondemand/ (accessed on 10/08/2011)

OFCOM (2008). *Market Impact Assessment of the BBC's local video service*, http://stakeholders.ofcom.org.uk/binaries/research/tv-research/bbc-mias/mia-bbc-local-video.pdf (accessed on 10/08/2011)

Office of Fair Trading (2006). *The commercial use of public information (CUPI)*, http://www.oft.gov.uk/OFTwork/publications/publication-categories/reports/consumer-protection/oft861 (accessed on 17/04/2011)

Office of Public Sector Information (2007). *The United Kingdom Implementation of the European Directive on the Re -use of Public Sector Information- the first two years*. London: The National Archives

O'Hara, K. (2011). *Transparent Government, Not Transparent Citizens: A Report on Privacy and Transparency for the Cabinet Office*, http://eprints.ecs.soton.ac.uk/22769/ (accessed on 20/10/2011)

Ohlenburg, A. (2004). "Der neue Telekommunikationsdatenschutz - Eine Darstellung von Teil 7 Abschnitt 2 TKG". *Multimedia und Recht*, 431-440 [in German]

Onsrud, H.J. (1995). "Role of Law in impeding and facilitating the sharing of geographic information". In Onsrud, H.J.and Rushton, G. (ed.). *Sharing Geographic Information*. New Brunswick: Rutgers, 292-306

Onsrud, H., Campbell, J. and Van Loenen, B. (2010). "Interoperable Open Access Licenses for the Global Earth Observation System of Systems (GEOSS)". *International Journal of Spatial Data Infrastructures Research* 5, 194-215

O'Reilly, T. (2007). "What is Web 2.0: Design Patterns and Business Models for the Next Generation of Software". *Communications and Strategies* 65, 17-37

Oxera (1999). *The Economic Contribution of Ordnance Survey GB*, http://www.ordnancesurvey.co.uk/oswebsite/aboutus/reports/oxera/oxera.pdf (accessed on 30/05/2011)

Pediaditi, K., Athanasiou, S., Nedas, K., Grigoriou, E., and Stefanakis, K. (2010). *INSPIRE Transposition, Member State Report: Greece*. Athens: MEECC

Perritt Jr., H. (1995). "Should local governments sell local spatial databases through state monopolies?". *Jurimetrics Journal* 35, 449-469

Peters, J. and Vriesendorp, R. (2002). "Insolvency of public entities other than the State under Dutch law". *Electronic Journal of Comparative Law*, http://www.ejcl.org/64/art64-23.html (accessed on 15/04/2011)

Pira International (2000). *Commercial Exploitation of Europe's Public Sector Information. Final Report*, http://ec.europa.eu/information_society/policy/psi/docs/pdfs/pira_study/commercial_final_report.pdf (accessed on 11/05/2011)

Pomfret, K. (2010). "Onus of the owner". *GIS Development* 14(6), 38-40

Potsiou, C. et al. (2010). "Rapid Urbanization and Mega Cities: The Need for Spatial Information Management". *Conference papers of XXIV FIG International Congress*, http://www.fig.net/pub/monthly_articles/march_2010/march_2010_potsiou_etal.html (accessed on 18/10/2011)

Power of Information Taskforce (2009). *Power of Information Task Force Report*, http://webarchive.nationalarchives.gov.uk/20100413152047/http://poit.cabinetoffice.gov.uk/poit/wp-content/uploads/2009/03/poit-report-final-pdf.pdf (accessed on 18/10/2011)

Price Waterhouse Coopers (2010). *Economic Assessment of Spatial Data Pricing and Access – Stage 1 Report: Principles, Issues and Alternative Models*, http://www.anzlic.org.au/Latest+News/275.aspx (accessed on 18/10/2011)

Prosser, T. (2005). *The limits of competition law. Markets and Public Services*. Oxford: Oxford University Press

Queensland Spatial Information Office (2006). *Government Information and Open Content Licensing: An Access and Use Strategy*, http://eprints.qut.edu.au/32117/1/Stage%202%20Final%20Report%20PDF%20Format.pdf (accessed on 18/10/2011)

Radoslavov, S. and Thomass, B. (2010). "The public value test as a societal debate about the future of PSB: The case of the German ZDF". Paper presented at the Ripe Conference 2010 'Public Service Media After The Recessions', London, 9-11 September

Ramachandran, R. (2000). "Public Access to Indian Geographical Data". *Current Science*, 79(4), 450-467, www.iisc.ernet.in/currsci/aug252000/ramachandra.pdf (accessed on 21/10/2010)

Rao, M. and Sridhara, M.K.R. (2006). "Keeping up with Remote Sensing and GI advances – Policy and Legal Perspectives". *Space Policy* 22, 262-273

RAVI Bedrijvenplatform (2000). *Economische effecten van laagdrempelige beschikkingstelling van overheidsinformatie*. Amersfoort: RAVI Bedrijvenplatform [in Dutch]

Rosenberg, P. (1958). "Earth Satellite Photogrammetry". *Photogrammetry Engineering* 24(3), 353-260

Roth, P. and Rose, V. (2008). *Bellamy & Child European Community Law of Competition. 6th ed.* Oxford: Oxford University Press

Sauter, W. (2008). "Services of general economic interest and universal service in EU law". *European Law Review* 33(2), 167-193

Snehmani, M. and Mathur, P. (2003). "Restrictions on Supply of Spatial Information on Maps", http://citeseerx.ist.psu.edu/viewdoc/summary?doi=10.1.1.131.4651 (accessed on 18/10/2011)

Solomon, D., Webbe, S., and McGann, D. (2008). *The Right to Information: Reviewing Queensland's Freedom of Information Act*, http://www.qld.gov.au/about/rights-accountability/right-to-information/assets/solomon-report.pdf (accessed on 22/5/2010)

Spatial Technologies Industry Association (2001). *Phase 1 report: increase private sector awareness of, and enthusiastic participation in the National Spatial Data Infrastructure (NSDI)*, http://www.fgdc.gov/library/whitepapers-reports/sponsored-reports/stia/ (accessed on 26/08/2011)

Steemers, J. (2003). "Public Service Broadcasting Is Not Dead Yet: Strategies in the 21st Century". In Lowe, G. and Hujanen, T. (ed.). *Broadcasting and Convergence: New Articulations of the Public Service remit*. Göteborg: Nordicom, 123-136

Steidle, R. (2009). "Datenschutz bei Nutzung von Location Based Services im Unternehmen". *Multimedia und Recht*, 167-170 [in German]

Stewart, K. et al. (2006). "Impacts of Licence Choice and Organizational Sponsorship on User Interest and Development Activity in Open Source Software Projects". *Information Systems Research* 17(2), 126-144

Suter, T. (2008). "Setting public goals, defining public value, securing public support – the challenge, for Europe's PSBs". Presentation at the conference on 'Public service media in the digital age. Strasbourg, 17 July

Tam, S. (2009). "Informing the Nation – Open Access to Statistical Information in Australia". Paper presented to the United Nations Economic Commission for Europe Work Session on the Communication and Dissemination of Statistics, www.unece.org/stats/documents/ece/ces/ge.45/2009/wp.11.e.pdf (accessed on 18/10/2011)

Tanner, L. (2010). *Declaration of Open Government,* http://agimo.govspace.gov.au/2010/07/16/declaration-of-open-government/ (accessed on 18/10/2011)

Technical Chamber of Greece (2009). *Report on Monitoring the INSPIRE Directive: Proposal for the Implementation and Management of the INSPIRE Directive in Greece.* Athens: Technical Chamber of Greece

Teresi, L. (2005). "Données publiques et concurrence". *Concurrences* 4, 143-156 [in French]

Tracey, M. (1998). *The decline and fall of public service broadcasting.* Oxford: Clarendon Press

Trappel, J. (2008). "Online media within the public service realm? Reasons to include online into the public service mission". *Convergence. The International Journal of Research into New Media Technologies* 14(3), 313-322

Tsiavos, P. and Korn, N. (2009). *Case Studies for the Mapping of Content, Value and Rights across the Public Sector.* London: Strategic Content Alliance

United Kingdom Information Commissioner's Office (2006). *Guidance on the Privacy and Electronic Communications (EC Directive) Regulations 2003,* http://www.ico.gov.uk/for_organisations/privacy_and_electronic_communications/the_guide.aspx (accessed on 18/10/2011)

United Nations (s.d.). *Treaties and Principles on outer space,* http://www.oosa.unvienna.org/pdf/publications/STSPACE11E.pdf (accessed on 26/08/2010)

United Nations Economic Committee for Europe. (1998). *Convention on access to information, public participation in decision-making and access to justice in environmental matters,* http://www.unece.org/env/pp/documents/cep43e.pdf (accessed on 18/10/2011.

United Nations General Assembly (1986). *Principles relating to remote sensing of the Earth from space,* http://www.un.org/documents/ga/res/41/a41r065.htm (accessed on 26/08/2010)

Valcke, P. (2004). *Digitale diversiteit. Convergentie van media-, telecommunicatie- en Mededingingsrecht.* Brussel: Larcier [in Dutch]

Van Boxtel, R. (2000). *Naar optimale beschikbaarheid van overheidsinformatie.* Minister voor Grote Steden- en Integratiebeleid. Tweede Kamer, vergaderjaar 1999-2000, 26 387, nr. 7, https://zoek.officielebekendmakingen.nl/kst-26387-7.pdf (accessed on 18/10/2011) [in Dutch]

Van Den Bulck, H. (2010). "Making Ends Meet: Developing Ex Ante Assessments in Flanders". Presentation at the IES & IBBT-SMIT Conference Exporting The Public Value Test: Assessing EU Member States' evaluation of public broadcasters' new media services. Brussels, 25 May

Van Den Bulck, H. and Moe, H. (2010). "Public Service Media governance after the crisis: comparing assessment practices in Flanders and Norway". Paper Presented at the RIPE Conference 2010 'Public Media After The Recession'. London, 8-11 September

Van Eechoud, M. and Van der Wal, B. (2007). *Creative commons licensing for public sector information. Opportunities and pitfalls,* http://www.ivir.nl/publications/eechoud/CC_PublicSectorInformation_report_v3.pdf (accessed on 17/04/2011)

Van Loenen, B. et al. (2007*). Open toegankelijkheidsbeleid voor geo-informatie vergeleken: het gras leek groener dan het was.* Delft: Delft University of Technology [in Dutch]

Van Loenen, B. and Van Barneveld, D.W. (2010). "Implementing INSPIRE: the process towards the harmonization of licenses for public sector geographic information in the Netherlands". *Proceedings of the INSPIRE conference, Krakow,* http://inspire.jrc.ec.europa.eu/events/conferences/inspire_2010/abstracts/118.doc (accessed on 26/08/2011)

Van Loenen, B. and Welle Donker, F. (2010). "Towards a framework of standard licenses for geo-information?!". Presentation at the GSDI-12 Conference. Realising Spatially Enabled Societies. Singapore, www.gsdi.org/gsdiconf/gsdi12/slides/1.3a.pdf (accessed on 18/10/2011)

Victorian Parliament Economic Development and Infrastructure Committee (2009). *Inquiry into Improving Access to Victorian Public Sector Information and Data*, http://www.parliament.vic.gov.au/images/stories/committees/edic/access_to_PSI/EDIC_ACCESS_TO_PSI_REPORT_2009.pdf (accessed on 18/10/2011)

Vlachopoulos, S. (2007). *Transparency in the State's Actions and Personal Data Protection*. Athens: A.N.Sakoulas

Volman, Y. (2004). "Exploitation of Public Sector Information in the Context of the eEurope Action Plan". In Aichholzer, G. and Burkert, H (ed.). *Public Sector Information in the Digital Age. Between Markets, Public Management and Citizens' Rights*. Cheltenham: Edward Elgar Publishing Ltd, 96-107

Welle Donker, F., Van Loenen, B. and Zevenbergen, J. (2010). "Geo Shared licences: a base for better access to public sector geoinformation for value-added resellers in Europe". *Environment and Planning B: Planning and Design* 37, 326-343

Werle, R. (1998). "An Institutional Approach to Technology". *Science Studies* 11(1), 3-18

Wetenschappelijke Raad voor het Regeringsbeleid (2000). *Het borgen van publiek belang*. The Hague: Sdu Uitgevers [in Dutch]

Wheeler, M. (2010). "The European Union's Competition Directorate: State aids and public service broadcasting". In Iosifidis, P. (ed.). *Reinventing public service communication: European broadcasters and beyond*. Basingstoke: Pallgrave Macmillan, 49-62

Whish, R. (2009). *Competition law. 6th ed.* Oxford: Oxford University Press

White, J.C. (2003). *People, Not Places; A Policy Framework for Analyzing Location Privacy Issues*, http://www.epic.org/privacy/location/jwhitelocationprivacy.pdf (accessed on 23/06/2011)

Wiedemann, V. (2004). *Public service broadcasting, State aid and the Internet: emerging EU law*, http://www.ebu.ch/en/union/diffusion_on_line/pub_broadcasting/ (accessed on 05/08/2007)

World Trade Organisation (1947). *General Agreement on Tariffs and Trade*, http://www.wto.org/english/docs_e/legal_e/gatt47_01_e.htm (accessed on 18/07/2011)

World Trade Organisation (1995). *General Agreement on Trade in Services*, http://www.wto.org/english/docs_e/legal_e/26-gats.doc (accessed on 18/07/2011)